RESIDENTIAL PATTERN OF SUBURBS

RESIDENTIAL PATTERN OF SUBURBS

SMITA SENGUPTA

CONCEPT PUBLISHING COMPANY, NEW DELHI-110059

ISBN 81-7022-028-9

First Published 1987
Reprinted 2004

Published and Printed by
Ashok Kumar Mittal
Concept Publishing Company
A/15-16, Commercial Block, Mohan Garden
NEW DELHI-110059 (India)

Phones : 25351460, 25351794
Fax : +91-11-25357103
Email : publishing@conceptpub.com

CONTENTS

LIST OF FIGURES

LIST OF TABLES

PREFACE

This study attempts to understand the pattern of residential area that evolves in suburbs of Indian cities. The pattern is essentially a morphological form that is brought about by a complex interacting set of processes in an environment. Here, in the specific case study, the setting is suburban, in eight maujas of the western part of the Ahmedabad city, studied during 1961-81.

The objectives of this study are to understand the growth and development of suburbs, especially residential suburbs. This study is not just concerned with the morphological or structural pattern, but also with the attitude and behaviour of residents. The study also aims at searching for processes that are responsible for such a pattern.

A Socio-Demographic' model, which is represented by a three dimensional rectangular parallelopiped graph, has been evolved to classify the residential pattern of the study-area. This model measures the level of 'Modernity' of the suburban residents in their socio-economic status, behaviour and life style and housing characteristics. These three factors represent the three axes of the rectangular parallelopiped graph. The term 'Modern', used in this study, means freedom of thought and technological sophistication, whereas the term 'Traditional' means the belief in age-old traditional institutions of a particular point of time and space. Variables in 'Modernity' of these three factors give rise to a residential pattern of households and housing structure on space. In maujas, where all three factors are modern, the pattern of those maujas has been termed 'Modern', whereas maujas where three factors are traditional have been termed 'Traditional'. Some maujas, which have the combination of both modern and traditional factors, have 'Mixed' pattern.

The relevance of studying the residential pattern of suburbs lies in understanding the growth of suburbs - not only in terms of morphology but also in terms of composition and behaviour of residents and how people organise themselves on space and which factors control such a distribution of population. An analysis of processes of the pattern will hopefully throw light on the growth of Indian suburbs on which very little literature is available. The classification of suburbs by housing and households can be helpful in micro-level planning, by understanding the need of each area in terms of the composition of residents.

Various methodologies have been used to test hypotheses which have been formulated to bring out the objectives of the study. The 'Socio-Demographic' model has been represented in a graphical fashion rather than through statistical method, as the nature of the data is qualitativly related to the behaviour, perception and attitude of residents. Both primary and secondary data have been used in this study. Primary data of 1500 households, relating to sixty variables, have been collected through field survey. All primary data are qualitative in nature, whereas secondary data, collected from various sources, are of quantitative type.

The study area, i.e. the residential suburbs of Ahmedabad city, is situated along the western side of Ahmedabad city. The study area has eight maujas of which Vejalpur, Jodhpur, Ranip and Ghatlodiya are much urbanised with the growth of new houses while Memnagar, Vastrapur followed by Chandlodiya and Thaltej, are less urbanised. The level of urbanisation has been measured by the rate of conversion of agricultural land into non-agricultural uses. This rate of change depends on various factors, e.g. nearness to the city, land value, socio-economic status of the residents of the core village around which suburbs grow etc. The study area has mainly three types of residential structures, i.e. core villages which are the original residential structures, existing prior to the development of the suburbs, housing societies which have been built due to suburbanisation and rest includes government colony, hutments etc.

In a cross-section of north to south in the study area, low growth of residential land use, low value of land and low growth of population is seen in the northern area where there is less development of various infrastructural facilities. In the south with good connectivity and accessibility the percentage of the residential area is high along with high land value and high growth of population. The northern section has residents with traditional socio-economic status, behaviour and housing characteristics. The situation is reverse in the southern side. In the central part of the study area, residents have the combination of both the traditional and modern characteristics of both morphological and behavioural aspects. So, the 'Socio-Demographic' model has formulated a slope of modernity in the residential pattern from south to north. This model has brought out that behaviour and life style do not change as fast as socio-economic status or housing characteristics of residents. It means that people take longer time to change their behaviour and life style, but external factors like housing, education, income can be changed faster. So modern housing as a result of suburbanisation may not match with residents' behaviour which may be traditional.

The study area has various problems like mixed landuse, vacant land without being used for a long time, lack of proper infrastructure, dual social characteristics of housing societies and core villages which do not have proper sanitation, and sewerage system. Core villages can soon be converted into urban slums. So government must initiate a number of rigid actions to solve all these problems. Micro-level planning is possible by knowing the attitude and behaviour and need of residents in the suburbs.

ACKNOWLEDGEMENTS

I am highly indebted to Dr. (Mrs.) Anjana P. Desai for her invaluable guidance, inspiration and affectionate advice during my research work. I am thankful to Dr. M.C. Jaiswal of the Department of Statistics of Gujarat University who has willingly helped me in the application of some of the statistical techniques. My thanks to Dr. N.J. Chinay of Zoology Department of Gujarat University for many practical suggestions. I am grateful to the authorities of Census of India, Ahmedabad circle; Gujarat Town Planning and Valuation Department; Ahmedabad Urban Development Authority; Ahmedabad Municipal Transport Service; Land Survey Office; Taluka Development Officer's Office; Ahmedabad District Collector's Office; Village Panchayat Offices; who helped me in procuring the relevant data, maps and reports from their respective organisations.

I wish to express my thanks to the authorities of Indian Institute of Management, Ahmedabad; Gujarat Vidyapith; Sardar Patel Institute of Social and Economic Research; School of Planning; M.S. University, Baroda; who helped me by allowing me to use their libraries for references. I feel obliged to the officials and residents of the study area who have willingly co-operated with me and have provided me with data on village and personal information. I am also grateful to Dr. P.S. Desai for consultation on computer programming.

My sincere thanks are due to Shri A.K. Mitra, Cartographer, who took efforts to draw maps of my thesis in a short time. I am also immensely thankful to Mr. C.S. Ramamurthi but for whose accurate, systematic, neat and steady typing work it would have been difficult to submit this thesis in time.

I am also grateful to all the staff members and students of Geography Department. I am very thankful to Mrs. Harsha C. Shukla and Mr. Bhupendra Patel of the Department of Geography, Gujarat University and my numerous other well-wishers, who helped me in several ways.

Finally, I am grateful to the University Grant Commission for awarding me the Junior Research Fellowship in April 1983 which enabled me to bear my expenses in research work. Last but not the least, this work could not have been completed without the encouragement of my family, which should not be taken for granted.

Smita Sengupta
Department of Geography
Gujarat University
Ahmedabad-380009

ONE

INTRODUCTION

1. Definition and nature of suburbs

With the growth of urbanisation, rural areas are constantly being engulfed into urban landscape, and villages in the immediate vicinity of cities and towns acquire eventually an urban character. The study of cities is, therefore, incomplete without a proper appraisal of the location, characteristics and development of these peripheral settlements, which have a potentiality of urban growth.

The Latin word "Suburbium" - i.e. suburb, is used in a very broad sense. In the medieval period, the total area of a settlement effectively the town, apart from the fort, was being called 'Suburbium'. The term was also sometimes used for an extensive area around the town, including other smaller settlements. In modern time, the term 'suburbs' means the "transitional" zone, located between purely urban and purely rural areas around a city.

Many scholars of different disciplines have discussed this peripheral fringe region as suburbs, but no precise definition or characteristic has emerged as yet. The problem of definition arises from the characteristics of the fringe itself. Being a transitional area in different aspects, i.e. location, landuse, demographic composition, socio-economic status and behaviour of residents, various terms have been used to denote this area, viz: urban fringe, umland, hinterland, etc. These different terminologies have been used by different scholar, depending on the aspects of their study of the peripheral area.

The expansion of urban residential area in suburbs is a result of the increasing population pressure within the city due to inmigration from adjacent rural areas and various parts of the country. So, due to the high population pressure in the city, people start moving into adjacent fringe areas of the city to have spacious accommodation in peaceful localities. With the urbanisation of the countryside, the sharp division of landuse and life-style of residents between urban and rural areas are becoming less and less perceptible today. Due to development of technology, transport and communication systems, it is becoming difficult to draw the boundary where the 'urban' stops and the 'rural' begins. The contacts between the city and the country are varied and are getting multiplied. They are both direct through commuting and wholesale and retail trade, and indirect through radio and newspapers and television. There is also socio-economic interaction between urban and rural areas. It is the intensity of urban-rural interaction that determines the pattern of contimum or dichotomy in landuse, occupational structure, friction of distance, demographic characteristics, life-styles and social attitudes of residents. In short, the suburban area is transitional in nature and is under constant flux.

The change in the suburban landscape is two-fold: (a) change in structure and (b) change in behaviour of residents. The change in the type and pattern of housing structure of the new suburban residents is seen along with the traditional houses of the original residents in core villages, around which suburbs have developed. The landuse of suburbs also gets changed from purely rural, agricultural to partly non-agricultural, urban oriented one. The transitional nature of suburbs is also reflected in the life-style and behaviour of residents, i.e. in terms of occupation, interaction, material possession and attitude of residents of this region. Due to the presence of urban population in new suburban area, the life-style of duel nature is seen among residents. The new suburban population has modern household items compared to those of traditional type in core villages. Infrastructural facility of suburbs is also transitional in nature with new residential houses having sewerage system, water-tap, electricity, etc., while core villages in suburbs have neither any sewerage system nor any water tap in most of the cases. There is also much difference in demographic composition of the core villages and new suburban areas. The residential structure of core villages is caste oriented with relatively low income and education but the residential structure of new suburban areas is mainly class (income) oriented with higher level of socio-economic status.

The presence of these two groups of residents gives rise to a unique residential pattern mainly in the suburbs of non-western cities, having to some extent a continuum in spatial structure and landuse but a dichotomy in spatial behaviour of residents.

The term 'pattern' is related to a certain 'shape' or 'design'. Here, it refers to the manner in which various processes have given rise to a morphological design or a pattern of suburbs, that is arranged on space. Patterns may exist in their physical form. They may be observed in the arrangement of the houses, alignment of transport lines, layout of service centres and infrastructure. In their abstract form, patterns are derived and deduced from the interpretation of qualitative aspect of social status of residents, their life style, behaviour, perception and interaction.

The transitional and dual characteristics of suburbs of non-western cities is also reflected in the residential pattern of suburbs, as these areas are developing into a modernised society with many traditional characteristics. In fact, it is not very easy to define the concept of "modernity" which is a very dynamic concept and it changes not only with the place and time but also with intellectual status of the people. According to Webster's dictionary, the term "Modern" means, "practices to adopt the conditions or tastes of the present time" whereas, the term "Tradition" means, "practices to posterity, a convention established by habitual practice or a belief or practice thus handled down." The concept of modernity in non-western countries has been generated through western contact, which means anything western is modern. However, it is not correct to say that, everything involved with Western contact has contributed to the process of modernisation. Similarly, everything traditional should not be termed as "antimodern". The idea of modernity is related to an enlightened framework of mind and technological sophistication. But changes due to technology may not go hand in hand with changes in attitude and behaviour of people. It is quite possible that, external pattern of housing of non-western suburbs can change faster due to technological development, but the internal behavioural pattern of residents may take longer time to change. This type of dual residential pattern with residents having traditional behaviour in modern housing in suburbs need to be studied for the understanding of the process of growth and of social organisation of suburbs. The suburban pattern in the western countries is different in the sense that, the residents with modern attitude try to maintain a rural appearance by their houses externally, whereas it is reverse in case of non-western suburbs.

2. The growth and characteristics of suburbs in Western countries

In Western Europe and in America, during the world War II and afterwards, it was noticed that, due to industrialisation the city was extending at a fast rate and there had been changes in landuse. Though in recent years the agricultural land is changing to residential areas very fast, suburbs were also present in pre-industrial cities. Suburbs of pre-industrial cities were inhabited by poorer groups. In this location they were remote from the city-centre which was the main hub of socio-economic and political activities. Gradually, in the second half of the nineteenth century those people who could afford to bear the cost of a relatively long journey to work and who had the desire to have specious house in a peaceful environment, had moved to suburbs from the city centre. Along with the population, there was a shift of amenities and infrastructural facilities also. So in this period, sites for those activities which could not be reasonably located within urban area due to their requirement of large space, were also located in suburbs, e.g. warehouses, sewerage pumping stations, oil storage stations, slaughter of the western cities were situated relatively nearer to the city limit and to the main transport routes, and suburbanisation implies then a separation between home and work place which was located at the core-city.

In the twentieth century, there has been a decline of population at the city centre in western countries and an outmigration of richer and younger group to modern suburbs. This was associated with the growth of per capita income, improvement of transportation and communication, introduction of car and need of private space per person in a peaceful unpolluted environment. So in this period, in western cities, suburban developments are not closely seen nearer to the city limit or on main roads. As a result, the suburbs of western cities are zones where more scattered urban expansion is taking place in interior locations.

In this type of scattered suburban development, the pattern of daily activity of residents are less directed towards a particular city-centre. Journeys to the city-centre for employment, shopping are considerably reduced; instead new shopping centres, new suburban employment, and new foci for recreation have made the city-centre only one of many possible destinations in the Western countries.

The growth of dispersed urban residential suburbs has reached its peak in late sixties with the expansion of ownership of vacation home in the more distant environment of large cities. In Western countries, at least a section of the urban population can afford to own two houses due to greater leisure time, increased affluence and the widespread availability of cars. As a result, the suburbs can be thought of as an extension of upto one hundred miles, of late, from the metropolitan cities. This countryside is easily accessible by road for a weekend and summer visit. This observation is based on conditions in the United States, although, even there "working class suburbs" have been established. These are associated with the growth of suburban manufacturing employment. The inhabitants of these suburbs, however, are rich enough to buy their own houses.

Elsewhere, in other western countries, different types of social groups are found in the suburban location due to the involvement of government in the provision of housing. In Britain, the council estates are government sponsored urban growth, which have brought bluecollar working class to the suburb. But the inhabitants of these council houses are rarely the poorest in the community. They are employed in suburban factories in skilled or semi-skilled jobs, rather than in unskilled occupation. The poorest in the society cannot afford to move to the suburbs even with government subsidy because bluecollar jobs are available in the city centre. Nevertheless, the suburbs of Britain and Western European Countries have residents with various economic status than is usually the case in the United States.

Suburbs of Western countries have a typical residential pattern, which can be identified by the presence of low density single family houses with accessible roads and recreation facilities. Some of these houses have traditional housing structures which is purposely constructed, with all modern facilities.

3. The growth and development of suburbs in India

Unlike Western countries, life in Indian cities is based on city-centre. In India, until recently, the decentralisation of population and economic activities from the city-centre has remained low. The socio-economic reasons for people to live at city-centre are as follows:

The majority of the people in India belong to the low income group. They save time & cost on transportation when goods and services are available at one place, that is, at the city-centre. Sellers want to have shops at a central location so that the total number of buyers may not be divided into smaller suburban shopping centres. Residents also prefer the shopes at the city-centre from the point of view of low price, variety and freshness of goods. Social reasons are also responsible for such a landuse pattern. Residents have developed a strong sense of attachment for ancestral houses and they enjoy social status and social interaction with friends and relatives and hence they do not like to leave the core-city. As a result, commercial activities, light industries are also found together with residential landuse at the city-centre.

Since the Second World War or so, the city limit in India has started to progress externally. This process of growth of residential suburbs in Indian cities is not exactly the same as that of Western cities in all respects.

In Western situation, there has been a decline of population at the city-centre and an outmigration of richer section of population is responsible for the growth of suburbs. This was accentuated with the growth of per capita income, transportation and communication, introduction of car and need of private space per person. In India, also richer and younger population have outmigrated to the suburb for want of more personal space and peaceful environment. But unlike the Western situation, a large number of rich people is still living in their ancestral houses at the city-centre, along with the incoming unskilled, migrated, rural population of low income group. So both the city-core and suburbs of Indian cities are inhabited by rich and poor sections of population and it creates a double standard of life-style in these areas.

In Indian suburbs, new suburban residents consist of the people from various parts of the city and not just from the core area. Here two-step migration is noticeable. In the first stage, inmigration of people from different parts of the state and the country takes place to the various parts of the city. Then in the second stage, most of these prople, along with some people from the core-city, migrate to suburbs by building own houses or getting house at cheaper rent. In short, the process of suburbanisation is not direct; people change their residence a couple of times before they settle down in the suburb in the form of housing societies.

Unlike the Western situation, the suburb in India is an extension of the continuous built up area of the city. This suburb includes the original villages which were situated outside the city limit, and also agricultural land. People living in these villages are engaged in primary activities, but when the demand of these rural land for non-agricultural residential purposes is increased, then these villagers sell their land to land developers or individual buyers, who are either outmigrating population from the city centre or from various parts of the city. In the suburbs of Indian cities, land located nearer to the city limit and transportation routes has more demand due to accessibility. So the richer section of people, who want to migrate in the suburb first, has settled near the city limit, then along the transport line and also around core villages which have people of higher socio-economic status. The people of the middle income group who migrate later in suburbs, generally settle in the interior where land value is less and around villages with people of low caste or near industrial centres. The low demand of land away from the city limit is mainly due to the low development of infrastructural facilities like, transport, shopping, education etc. in the interior of the suburb.

The people of the suburb depend on the city-centre for their daily needs for their economic and social interactions and hence the demand of accessible land is more. Some of the new residents of suburbs have their friends and relatives in the core-city, so they commute frequently for social interactions to the core of the city to meet friends and relatives. But this journey to the city is not a very easy job, because due to fewer private vehicles, suburban population depends mainly on insufficient public transport system for commuting to the different parts of the city.

In suburbs, where government housing schemes are organised by income, viz: high, middle and low, residents get automatically segregated by economic class. But in private housing societies, sometimes groups of the community by language or religion get together to have their own area of community life. Three levels of interaction pattern are seen: interaction within the core village, interaction between the suburb and the core city and interaction among the suburban population.

Unlike Western suburbs, in Indian suburbs, original core villages are still existing. The people of these core villages still have their traditional culture but they now face the crisis of identity due to the infiltration of modern urbanite population. So the pattern of the suburb in India is gradually becoming complex as it is changing to an urban area from a rural environment. The people, especially the older generation of the original villages like to keep traditional identity as much as possible regarding their occupation behaviour, life-style, interaction pattern and housing structure. There is very little interaction between the people of core villages and those new suburbs. These groups have much difference regarding socio-economic status, behaviour, life-style, material possession, interaction pattern and housing structure, which have given rise to a dual residential pattern in Indian suburbs.

In India, suburbanisation is related to the shift of population in newly developed areas with new housing structure and new pattern of interaction. But even in modern suburban societies there are certain ways of traditional life-style, e.g. caste system, which has regulated Indian rural life for several centuries and which plays an important role in the growth of the process of suburbanisation. Not only caste system, but various traditional beliefs (superstitions), rigid food habits, ageold social values related to the low level of female education, early marriage, belief in fatalism are sensitive issues both in rural and urban life-style in India. So like other non-Western countries, external pattern of Indian suburbs has changed faster due to technological development. But the traditional behaviour of residents has not changed to that extent. So one can expect to find duality in modern housing societies with traditional households in the suburbs of the Indian cities.

4. Objectives of the Study

The main objective of this study is to understand the residential pattern of the suburbs of Ahmedabad city during 1961-1981. But this residential pattern, is not just the morphological or structural pattern of the suburbs but also of the social organisation or pattern as depicted in the attitude and behaviour of residents. This study also aims at searching for processes that are responsible for such a pattern. This study has focussed on the suburbs of Ahmedabad in terms of processes of growth, morphological changes, demographic composition, and the behaviour of suburban residents in order to find out a composite residential pattern of the suburbs.

The study has demarcated residential suburbs from the total suburbs of Ahmedabad city, based mainly on landuse, demographic composition and connectivity (Figure 1.1.). The study area is located on the western side of the river Sabarmati.

Total suburbs of purely residential area of Ahmedabad city have been divided into four groups. Two subdivisions of this primary division are further subdivided into three groups and two of which have been grouped into 'Residential Suburb' (methods of divisions have been discussed in Chapter-3).

The residential pattern of suburbs is a complex phenomenon of interacting processes of various morphological, demographic and behavioural aspects of residents (Figure 1.2.). These factors are analysed below to understand the growth of a pattern. The morphological factor is the combination of some physical variables like landuse and housing structure on space and the behaviour factor is the combination of perception, interaction attitude and life-style of population living in the suburbs. Both morphological and behavi-

Fig. 1.1

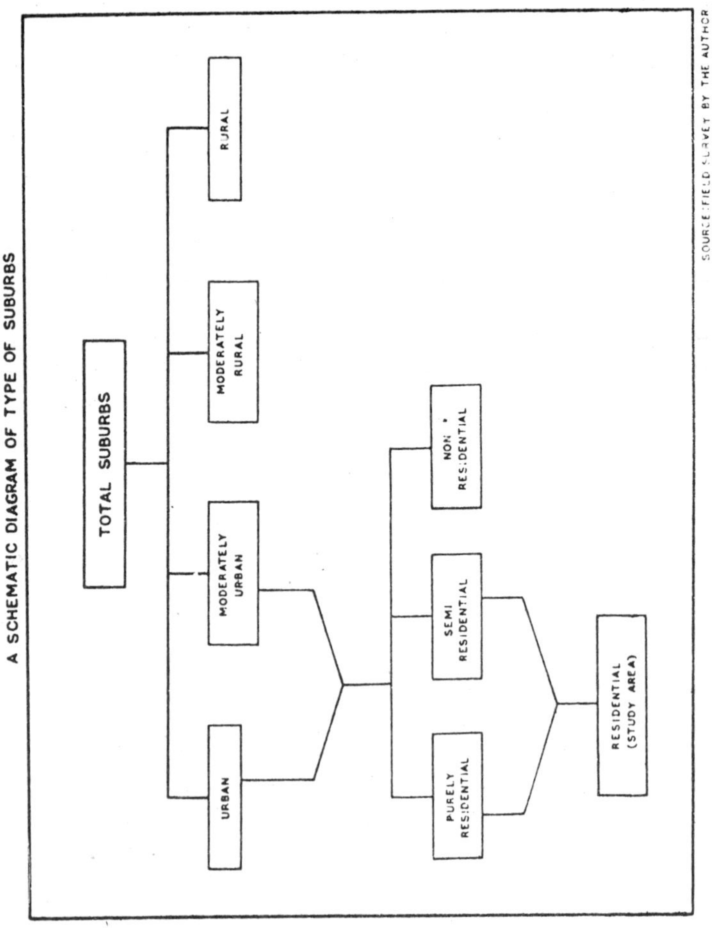
A SCHEMATIC DIAGRAM OF TYPE OF SUBURBS
TOTAL SUBURBS
URBAN
MODERATELY URBAN
MODERATELY RURAL
RURAL
PURELY RESIDENTIAL
SEMI RESIDENTIAL
NON RESIDENTIAL
RESIDENTIAL (STUDY AREA)
SOURCE: FIELD SURVEY BY THE AUTHOR

Fig. 1.2

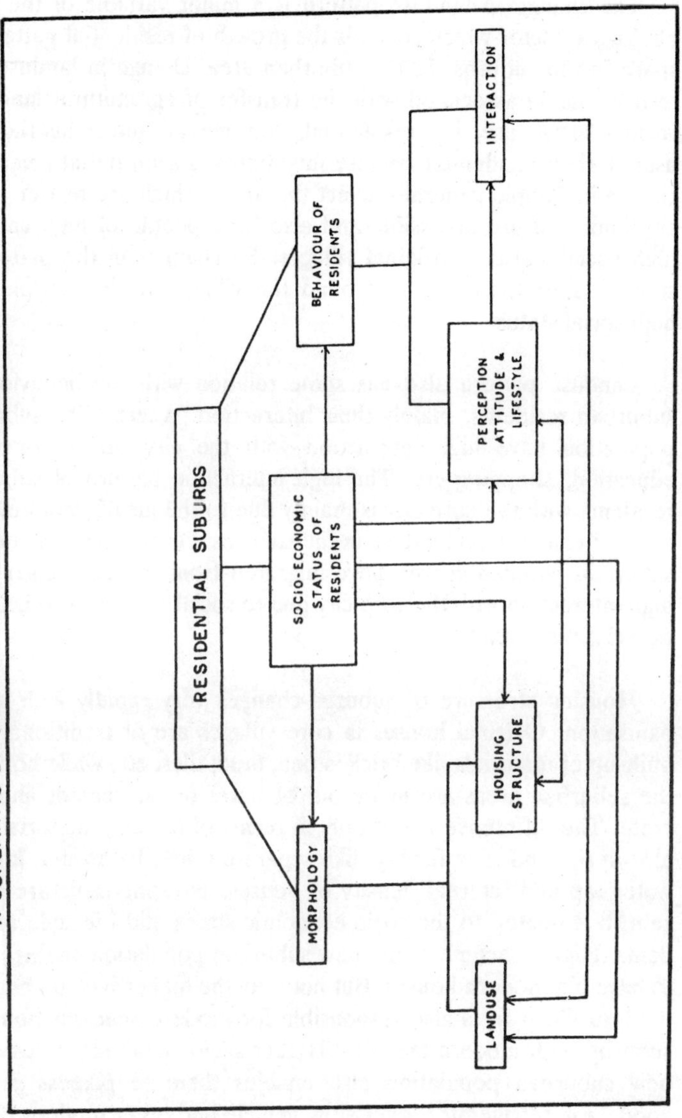

A SCHEMATIC DIAGRAM OF THE PATTERN OF RESIDENTIAL SUBURBS

oural factors have relationship with socio-economic status of residents, i.e. their level of income, education and occupation.

The change in landuse pattern is a major variable of the morphological factor which controls the growth of residential pattern on space in the suburbs. In the suburban area, change in landuse pattern is mainly associated with the transfer of agricultural land into various urban use, i.e. residential, commercial and industrial land uses. Urban residential landuse in suburbs is seen initially near the city limit. People generally select the areas which are nearer to the city limit and are accessible and also have people of high caste or high social status in original villages. So changes in the pattern of suburbs start taking place around the villages having people with high social status.

Landuse pattern also has some relation with the behaviour of suburban residents, mainly their interaction pattern. The suburban population have high interaction with the city centre for work, education, shopping etc. The high interaction pattern of suburban residents with the core city is mainly due to the insufficient development of commercial landuse in the suburbs. On the other hand, new suburban population who have migrated from the city centre have high interaction with the core city due to social reasons also i.e. visiting relatives, friends etc.

Housing structure of suburbs changes very rapidly with suburbanisation. Original houses in core villages are of traditional type, built-up of materials like brick, stone, mud, tiles, etc. while houses in the suburban areas are made out of burnt brick, cement and concrete. The difference is not only in terms of building materials, but also in size; housing facility, like separate toilet, bathroom, kitchen, water tap and internal density of houses. Housing structure of the suburb is related to the socio-economic status and life-style of residents. Higher income of the new suburban population enables them to have big modern houses. But not only the higher income but their modern life-style is also responsible for modern spacious houses in suburbs, with modern facilities. Higher socio-economic status of the new suburban population also enables them to possess modern goods, e.g. refrigerator, television, vehicle and other modern electrical gadgets. Though the life style, i.e. possession of sophisticated goods and housing structure of suburban population, may have changed at a very fast rate, their behaviour and culture do not necessarily change in the same rate. This aspect need thorough investigation. So these uneven changes of physical and behavioural status of the suburban population create a unique mixed type of residential pattern in suburbs.

This study intends to analyse some aspects of the above-mentioned residential pattern and residential behaviour along with the morphological changes of the residential suburbs of Ahmedabad city during 1961-81. It is hoped that a general picture of Indian suburbs can be evolved through this case study. The main objectives of this study can be specified as follows:

(a) To demarcate the suburban zone with special reference to residential suburbs;
(b) To find out the step-wise migration of suburban population, by evolving a generalised model;
(c) To analyse various aspects of suburban landuse, demographic composition and distribution of infrastructural facilities which are responsible for the morphological growth of suburbs;
(d) To find out characteristics of suburban population regarding their socio-economic status, behaviour and life-style and housing characteristics;
(e) To evolve a 'Socio-Demographic' model which will combine the socio-economic status, behaviour, life-style and housing characteristics of suburban population into a three dimensional parallelopiped which will help in the classification of suburbs on a micro level.

This model provides an analytical framework to understand and to classify the residential pattern of structure and behaviour of the study area. The main purpose of this model is to measure the level of modernity of suburban population and their organisation on space. Suburbanisation is not modernisation in all respects, especially in India, where people have faith in ageold traditional belief and customs. So the main significance of this model is to bring out a complex combination of a pattern of traditional values of life in modern suburbs.

5. Relevance of the Study of the residential pattern of suburbs

The study area comprises of eight maujas (revenue villages) adjoining the western periphery of Ahmedabad city. This area has experienced an enormous growth of population, a decrease of agricultural land and a consequent increase of non-agricultural land in last two decades (1961-1981). Though most of these maujas (except one) are just across the city limit, all of them have not grown at the same rate. This growth and development is influenced by various morphological factors, socio-economic status and behaviour of residents.

The study helps to understand the growth and development of suburbs, as follows:

(i) It has helped to throw light on the problem of the definition and demarcation of suburbs:

(ii) This study has helped to understand the process of growth of suburbs through migration in various stages. It also gives an idea of the composition and nature of suburban population;

(iii) This study indicates the uneven and unplanned growth of suburbs which will be parts of the city in future and it has also given suggestions to solve these problems, e.g. government control over land transfer and also the control of land value;

(iv) The inadequate facility of infrastructure of the suburbs also have been indicated in this study. So this will help planners to plan for the future development of infrastructure in suburban areas;

(v) The study has shown that there are differences regarding socio-economic status, behaviour and life-style and housing characteristics, not only within the maujas, i.e. core villages and new suburban housing societies but also between the maujas, with the help of the 'Socio-Demographic' model. So a proper planning is needed to remove variations between core villages and housing societies so that core villages do not become urban slums in future;

(vi) This study evolves the concept of the 'Modernity Index' by which it classifies the suburbs by housing and households in terms of structure and behaviour. The concept of the 'Modernity Index' is the main contribution of this study and it

> shows that, suburbanisation is not associated with modernity in every aspects of life. The term, 'Modernity' used in this study, is to understand the level of changes in social, economic and behavioural aspects and how these changes are associated with industrialisation and westernisation. This study concludes with the help of the model that the physical changes of the suburbs are more rapid than the behavioural changes of residents.

The relevance of studying this residential pattern of suburbs lies in understanding the growth of suburbs in Indian cities, not only in terms of composition and behaviour of residents and how people organise themselves on space but also which factors control such a distribution of population. The classification of suburbs by housing and households can be helpful in planning the needs of residents on a micro level, so that the variation in housing and infrastructural facilities over space can be minimised. The findings of this study will be useful to planners and decision makers who can develop the suburb in a comprehensive manner by controlling the development of landuse, by providing necessary amenities to residents - both to those of the core villages and to those of new suburbs by understanding their behaviour, attitude and life-style.

TWO

LITERATURE SURVEY

The analysis of suburbs in various ways has been the objective of both theoretical and empirical work in the post World War era. Our concern here is with the choice of an appropriate conceptual framework for the analysis of this study.

The extent, characteristics and problem of suburbs vary from one city to another. To analyse it, a number of classical studies have been used not only by geographers but also by sociologists and economists. The conceptual and analytical framework has been developed by these scholars. This section will be devoted for the survey of literature related to the various aspects of the pattern of suburbs. This survey will provide us a valuable insight into the different aspects of suburbs that may later be used to generate a model of residential pattern of suburbs which is relevant to the Indian situation.

1. Problems of delimitation and definition of suburbs

The subsurb is not an easily defined geographical area that begins and ends at a certain distance from a city centre. It is rather a transitional area where dominant functional use of land is not clear. Therefore, rapidly growing urban centres, having a peripheral suburban region around them, was called "Rural-Urban fringe" by Von Thiinen (1826). The term fringe suggests a transitional area between the rural and urban zone having the characteristics of both the areas. The confusion in delimitation results from various case studies which are based on different aims and the context of research.

Hawlay (1956) on the basis of 'growth pole' as a conceptual framework of urban influence, delimited an area of 25 miles where growth of population was urban oriented. At the same time, he also delimited the primary commuting areas of daily interaction with the city in terms of employment, shopping etc. upto fifteen miles from the city centre which may be called an area of urban influence. Berry, Fox, Garrison (1966) delimited the suburban area by the distance used by daily commuters for work, immediate shopping and social trips upto twentyfive miles. Israd and Whitney (1949) have approximately estimated twenty miles radius as the commuting range of metropolis by road. Queen and Carpenter (1953) have studied area outside the Melbourne city. Similarly Duncan and Reiss (1956) have used the variable of suburban population in urbanised area, lying outside Chicago city. Martine (1957), Rohrer and Hirzel (1957), Golbesge (1960), Reinemann (1960), Dindrwes and Eshlemann (1963), Pahl (1965), Goldsmith and Lee (1966) and Johnson (1966) had arbitrarily taken an area of five miles radius from Melbourne C.B.D. and did not consider physical and cultural factors to demarcate suburbs. He had not given thought to the lines of transport along which the fringe belt is bound to fluctuate.

There is also much diversity of definition of suburbs. Kurtz and Flitcher (1958) had illustrated the problems to differentiate between 'fringe' and 'suburb'. Wissink (1961) in an extensive survey of the fringe areas of American cities identified 'fringe', 'suburbs', 'pseudo suburbs', 'satellities, 'pseudo satellites'. Schnore (1952) distinguished 'urban fringe' and'external fringe'. Reinemann (1960) distinguished 'suburban fringe zone' and 'outlying adjacent zone'. Wissink (1962) had identified 'inner' and 'outer' fringe areas. Duncan and Reiss (1956) had postulated 'rural non-farm' and 'rural farm' sub-divisions of suburbs, while Myres and Beegle (1947) were able to distinguish between a 'true fringe'. 'partial fringe' and 'adjacent rural township'. Whitehand (1967) had given the terms 'inner fringe belt' (IFB), 'middle fringe belt' (M.F.B) and 'outer fringe belt' (O.F.B). He suggested that, inner fringe belt was surrounding the inner city, while adjacent builtup area was known as middle fringe belt and it was followed by outer fringe belt, which was consisting of countryside. Oosthwizen (1969) proposed various terminology, for example, 'suburban area', 'rural suburb' and 'quasi-urban area' or 'quasi-urban township'. According to him, suburban areas are the residential areas situated outside but adjoining the municipal boundaries. These areas have no urban local administration but have distinct urban character and are closely linked to the city economically, socially and there was also a considerable amount of daily commuting between the suburban area and municipal town. The 'quasi-urban township' meant villages with clusters of population which were situated away from metropolitan areas.

Pryor solved the problem of various differentiations of an urban fringe by sub-dividing the suburbs quantitatively into 'urban fringe' and 'rural fringe'. He delimited the area on the basis of transition in landuse and social and demographic characteristics lying between continuously built up urban and rural areas.

Several authors in their studies have attempted to delimit suburbs based on various characteristics, namely social, demographic and landuse pattern. None of these above-mentioned characteristics are solely helpful to demarcate suburbs in a comprehensive manner. Due to the modern methods of communication and good road system, there is very little distinction between urban and rural areas, regarding life-style of residents, their occupation in developed Western countries. Therefore, in developed countries, landuse rather than social or employment factors, would provide a simpler and more reliable guide to demarcate suburbs.

2. Sampling techniques for demarcating suburbs

Most of the studies of suburbs deal with a portion of suburban population which are selected by different sampling techniques that are suitable for the particular study. Andrews (1946) took 595 cases in Janesville, Wisconsin, but he did not state the source of data. Gist (1952) interviewed 460 families from 1800 residents of the fringe area in Columbia, Missouri, but did not indicate any sampling procedure. Faust (1942 selected three voting precincts outside Eugene, Oregon, on the basis of representative parishes in the suburb of Sydney. Firey (1946) interviewed about 100 persons in three widely separated fringe-neighbourhoods of Flint, Michigan. Sample was selected at random to represent all types of dwellings and only 84%

of the results were handled statistically. Press and Hein (1962) studied a non-random sample of 107 out of 167 farm families in Alaiedon, Michigen. Pahl (1965) had selected three parishes in Hertfordshire, London, by a process of elimination. Other research workers have used more objective selection techniques. Mckain and Burnight (1953) analysed census data for five townships which are peripheral to the Hartford metropolitan area, because these townships had all the four criteria selected by them for their study. Zimmer and Mawley (1958) studied local government as perceived by fringe residents. Their study was based on 413 interviews from 2% sample of households. Andrews and Eshleman (1963) interviewed 303 households heads in the fringe of Ohio, using a multistage sampling system based first on an area and second on population.

Therefore, to select sample size, both random sampling and stratified sampling methods were used. There are various factors which control the choice of a particular sample. It may be physical, economic, social and cultural factors of towns and suburbs. In random sampling, there is justification in comparing cases and statistical analysis based on it has a sound foundation. But in case of purposively selected samples, the result can be justified under particular conditions and under certain analytical procedures and comparisons are not always appropriate.

3. The characteristics of suburbs

The suburban growth in the postwar period has encouraged social scientists to pay attention to the study of landuse and functions of suburbs. The suburb of an urban complex has a pattern and characteristics depending upon the physiography and transportation facilities of the area and also on the characteristics of urban complex itself.

Present idea about characteristics of the suburb originated from an article, published by Wehrwin (1942), which has now been regarded as one of the basic papers in this field of study. In his investigations of Indianapolis, he recognised the growth of railways as the main cause of decentralisation of urban population and of changing of landuse from farm to non-farm ones. Such changes were enhanced by the introduction of car and metalled roads which provided a fast system of mass transportation. Von Thiinen (1826) in his major work. "The Isolated State'postulated the structure of a simple agricultural zone but that had been modified by railways. Whenever railways, highways were present, a city's residential, commercial establishments and industrial plants developed in a linear fashion along with their routes.

Examining closely the urbanised part of suburbs, Wehrwin identified its important characteristics, mainly the nature of its residential growth. Due to the development of the transport system between city centre and suburbs, urban areas expanded quickly in American cities. Speculators and land developers rushed to suburbs to buy large areas of agricultural land which they left vacant initially and would sell gradually at a higher price. Therefore, in addition to the areas of built up land in the suburbs, there are speculated land which are kept vacant purposely.

The second characteristic feature of suburbs was indicated as the location of recreation land and institutions. Many of the public utility services which require space were located in suburbs though they were serving the needs of residents of cities, for example, radio station, sewerage disposal plants, airport, cemetry, slaughter house, wholesale oil storage, various heavy and large scale industries, junkyards. Some of them are necessary but unwanted landuse in the city, while some other activities have some restriction of location within the city limit. So these landuse are located in the suburb in order to escape rules of land control. In countryside, rural authorities have relatively less control over these matters of landuse.

Wehrwin's pioneering study had been followed by many others, focussing upon the characteristics of suburbs. Balk in his work on Worcester, Massachusettes, have a detailed description of suburbs. Andrews (1942) had attempted to identify the general landuse characteristics. Dickinson (1956) identified suburbs and analysed its functions. According to him, suburbs were invaded by the urban land uses by the extension of housing with buildings along the main arterial roads and by location of new factories which are excluded for one reason or another from the compact urban area, golf courses, waterworks etc. Pryor (1966) characterised the rural-urban fringe by the incomplete availability of utility services, inadequate network of public transport and a relatively high ownership ratio of residential houses. The majority of the work places and the area of retail purchase of residents of suburbs were seen in the city itself. Kurtz and Smith had looked, in detail, at the social characteristics of suburbs which had rural background with urban residents.

Literature dealing with characteristics of suburbs are diverse. But main characteristics of suburbs as summarised from various research work, are rapid growth of residential and institutional landuse but suburbs have less development regarding infrastructural facilities.

4. Process related to the growth of suburbs

One of the initial studies in this topic is by Golledge, who, in 1960, published the results of his investigations of Sydney's suburb. This study is also the first one which dealt with a fringe area lying out of Western Europe and North America. The work was conducted on two periods - one before landuse controls were introduced, the other following the introduction of fairly strictlaw regarding the maintenance of the green belt. According to him, urban suburbs have seven major characteristics: constantly changing landuse pattern, small farm, intensive crop production, mobile and low and moderately dense services and public utility, rapid residential expansion, common speculative sub-division of buildings. He examined each characteristic with respect to Sydney's suburb.

Sydney's green belt, like that of any other cities in America and Western Europe, had an intensive effect upon the fringe. Its purpose was to act as a buffer zone between the city and rural areas and to provide space for recreation and other pursuits, to reserve areas for the use of government and to create an area around city where desirable type of agriculture, recreation and amenities could be maintained.

The second study which throws light on processes of suburbs is by Pahl. His work largely focusses on social character of suburbs of London and difficulties of defining suburbs in social terms of demarcating the social identity of suburbs but he did not point out the urban influence in social changes of rural and semi-rural areas. He recognises four main processes. i.e. tendency towards social and spatial segregation, selective immigration, changing character of commuting and collapse of geographical and social hierarchies.

These two studies of Golledge and Pahl dealt with processes within urban suburbs but their standpoints were widely different. Golledge explained the growth of suburbs through physical processes while Pahl has a social approaches an explanation for suburban growth.

5. Problems of Suburbs

The mixture of urban and rural landuse and of urban-rural society in suburbs creates a complex atmosphere which generates various problems. David Thomas categorised the problems of suburbs in the following divisions:

(i) There are those problems which stem from the scattered residential and commercial development in unplanned suburbs;
(ii) Difficulties arise in suburbs from the inter-mixture of agricultural and non-agricultural landuses;
(iii) Problem generates from the high costs of services to scattered settlements. Costs are not high when compared with truly rural areas but they are quite high in comparison with nearby urban areas;
(iv) There are difficulties which follow from the inter-mixture of different social groups, some with urban based behaviour and others with rural based attitude of life.

6. Landvalue of the city and suburbs

The study of landvalue of an area is very much essential to understand the growth and development of that area. The value of a particular land is controlled by many factors, and it is very much influenced by the land use of that area. Not only land use but many other factors control landvalue of an area. Many economists have worked on land value of cities and their suburbs and some of them are discussed below to understand the different factors which control landvalue of an area and also the relation between landvalues and land uses.

One of the earliest work of landvalue was done by Hard (1908). According to him, the economic rent of cities is based on superiority of location. So the rent at the most accessible point is high which will increase the landvalue of that point. This accessibility depends on nearness to the Central Business District of a city which has the maximum landvalue and this landvalue declines sharply within a short distance of the Central Business District.

Haig (1926) determined that, if there be no transport cost then all locations would have gotten the same land value. According to Ratcliff (1949), landuse is determined by the rentpaying ability of various users in various locations. He drew some figures to derive city structure from broad category of landuse by their ability to pay rent and distance from most accessible point. He concluded that retail users with high paying ability could afford the greatest accessible area. Then it is followed by offices. Then comes multiple family housing complex, for example, apartments. Then follows single family housing and lastly agricultural land.

Alonso (1960) added two more variables 'quality of land and amount of disposable income' devoted to land and transport cost. If the amount of service is held constant and the relationship between quantity of land and distance considered, then the price of land decreases with the increasing distance from the city centre while quantity of consumption of land increases. But if transport cost with increasing distance increases then the amount of land to be purchased decreases.

Yeates (1968) in his study of land value in Chicago tested the influence of number of variables including distance from the Central Business District, distance from the nearest regional shopping centre and distance from nearest elevated subway system. All of these variables proved significant in a distance-decay manner, but supported the proposition that, location of a land is important in influencing its value, that is to say, the greater is the distance from the transportation system, the lower is the land value.

Brigham (1965) explained that, since land supply is fixed, land value in an urban area is determined by the demand for space. This demand function, for any site is a function for sites accessibility, level of amenity, topography and certain qualitative phenomena that may be considered, for example, image of the area and value of the land in non-urban use. According to Northam (1975) in the case of residential land values, accessibility to places of employment, distance from the central city are the most significant variables. Relative location with respect to other activities, influences land values. There are, of course, physical influences on land values though it can be modified or overcome by reducing the effect of the constraints.

So the land value of a city is the highest in the C.B.D. which is most accessible. The land value decreases in an outward direction from this most accessible centre, outwardly in suburbs. Value of land is also controlled by the nature of its use. Residential land value of a suburban area varies with respect to its location to other activities, for example, transport routes, shopping centre, place of employment etc. All of these above mentioned studies are based on Western situation which is somewhat different from the Indian situation. In India, residential land value depends on accessibility, distance from the city and also on the caste or religion of residents, while in Western countries the racial factor is given importance in determining the quality and value of a neighbourhood. In short, the study of land value can indirectly indicate the socio-economic status of residents. The rich community will be located near the areas having more facilities and the land value of these areas is obviously high. The case is reverse for the areas with low land value.

7. Different aspects of Perception Studies

The above mentioned literature is based on the structural and morphological aspects of suburbs. But the present study will also discuss the spatial distribution of pattern as depicted in the behaviour of suburban residents. So to understand different behavioural attitude of residents, it is necessary to study different aspects of perception.

Since 1960 onwards the environmental perception was recognised as an important concept. But in the writing of Sauer, Wright and Krik, the idea that the people behave in the environment not on the basis of objective knowledge but in terms of subjective interpretation, has appeared much earlier in the late 1940's and early 1950's.

This concept of 'man-environment-relationship' has been taken up in several research projects, theoretically and empirically at various levels. Mental image and the process of image building have been two main aspects of research. The majority of the study on mental image are concerned with human behaviour.

The perception of neighbourhood by the residents had been studied by Lee (1963). According to him, residents generally get attached to their own neighbourhood in ten years. According to H.J. Klein (1967), people have different opinion about central business district. He gave more importance to the image than the behaviour of population. Downs studied the perception of shopping centres in Bristol. He found that the residents have a better idea of downtown shopping centres than those of neighbourhood ones. Lynch's (1960) work based on psychological concepts is an earliest and pioneer work on image study. Mental images or the inhabitants of Boston, New Jersey and Los Angeles were traced, based on different responses like street interviews, verbal descriptions and sketch maps. The concept of 'residential desirability' was used by Gould and White in a study of school children in U.K., U.S.A.,Ghana and Nigeria.

8. Empirical Work on Indian Situation

Articles and books which were published mainly based on the studies in North America, Western Europe and Australia do not appear to be totally applicable in the Indian context. Indian cities are still core oriented, therefore it is not necessary that mainly rich people always migrate to suburbs. Suburbs of Indian cities, unlike Western countries, have grown without much infrastructural or residentiai facilities. So suburbs of Indian cities have to be studied in different perspective.

In India, sometimes suburbs are studied as a part of the whole metropolitan area. Kar (1963) attempted to investigate the characteristics and norms of the metropolitan economy in and around Calcutta. Bagchi (1966) studies Howrah conurbation as an urban sprawl of Calcutta Metropolitan region. Singh (1967) has described suburbs as the rural land with urban phenomena. Chattopadhyaya (1967) studied the transport nodes of a part of Calcutta Metropolitan region which is a part of suburbs of Calcutta. Alam (1972) used some principal and reflective elements, for example, suburban transport service, commuting areas of workers of factories, retailing, water supply, percentage of non-agricultural workers, electricity consumption, growth rate and density of population etc. to study patterns of development of suburbs of Hyderabad. Jadav and Kulkarni (1967) studied the landuse planning and its problems of suburbs of Poona. Nangia (1976) studied various aspects of the settlement pattern of Delhi Metropolitan region. Bor Gowda and Mahadev (1977, 1981, 1982) studied the changing land use pattern, growth and, development of co-operative housing societies, process and policies of landuse conversion in the suburb. Ramchandran and Srivastava (1974) suggested a model of 'Rural-Urban Fringe Sequence' while studying the suburbs of Delhi. They formulated the model showing the stages of change in the suburb. Sinha (1980) studied the impact of urban influences in the rural-urban fringe of Patna and the change in the cropping pattern in response to the demand from the urban market. Hussain (1980) emphasized on haphazard urbanisation by urban encroachment on rural lands of Muzaffarnagar. Karba (1980) identified a set of appropriate action programme for the improvement of Indian suburbs with reference to Tajganj, a small neighbourhood of Agra. Kumar (1980)made a study of suburbs of Bhopal and suggested the relevant landuse planning. She presented a model which is different from Von Thunen's model by examining the trend of diffusion of agricultural innovation.

Phadke and Sita (1981) studied various aspects of Bombay Metropolitan Region. They delineated the suburbs by using 'Boundary Girdle Technique' and the'Synthetic Index Technique'. They also studied economic characteristics of metropolitan region. Gopi (1981) suggested a model of the development process of suburbs of Hyderabad. Wadhva (1982) studied the land value and landuse of suburbs of Ahmedabad. Some attempts have been made by the Indian geographers to study the different aspects of the perception of the environment.

The problems chosen for the study of environmental perception have been quite diverse. Desai (1980) has chosen the core city of Ahmedabad for the perception studies of urban environment. Her work, one of the earliest example of perception studies in Indian context, deals with the sensitivity and adaptability of the residents in different environmental problems and how residents evaluate their environmental quality. Kayastha (1980) has been engaged in studying the perception of flood hazards in Eastern Uttar Pradesh. Saha (1982) has studied the role of cognitive processes in perceiving the elements of a part of Calcutta city. The human adjustment to power crises in Calcutta has been studied by Nag (1980). Studies related to the perception of environment in urban areas is becoming popular in India.

From the above mentioned literature it has been found that a great deal of work has been done about different aspects of city and suburbs. In short, Pryor, Andrews, Ramchandran, Sinha and Gopi analysed the problems of delimitation of suburbs. Wehrwin, Pryor, Kar, Singh, Jadav and Kulkarni and many other authors have explained the characteristics and processes for the growth of suburbs, while many other studies brought out the problems of suburbs.The perception and interaction pattern of urban residents has been studied by Desai. These studies have helped to understand the different aspects of suburbs.

9. Contribution of the present study

The main thrust of this study is to find out the characteristics of residential suburbs in terms of morphological, socio-economic and behaviour of suburban residents. The study will concentrate on the type of people that have settled in suburbs - their socio-economic status, life-style and behaviour, interaction pattern, problems they face and their perception. No study has concentrated on the behavioural aspects of residents of suburbs. So far studies have been on the structural changes in suburbs. This study will look into both the structural and behavioural pattern of residential suburbs.

To begin with, this study has analysed the process of growth of suburbs in a longitudinal manner with the help of step-wise migration, tracing fifteen hundred individual migrants, movement for four locations. Then the movement was analysed in terms of origin-destination matrix stage-wise flow chart and a generalised spatial model of migration.

Further, the contribution of this study lies in the analysis of the spatial structure and the spatial behaviour of suburban residents in terms of levels of modernity in three major factors, i.e. their socio-economic status, behaviour and life-style and housing structure. A Socio-Demographic model has been proposed to understand the composite picture between residents and residential units, in above mentioned dimensions with the help of sixty variables.

This 'Socio-Demographic' model, which has been represented by three major factors, i.e. socio-economic status of residents, their behaviour and life-style and lastly their housing characteristics, helps us to classify the suburbs on a microlevel to understand which groups of residents are living at which localities. In other words, the process of suburbanisation is seen in modern houses with residents of modern ideas, or in modern houses with residents of traditional thought or in traditional houses with residents of traditional thought. This model is the main contribution of the present study which helps us to understand the relation between residents and residence and the spatial organisation of suburban population.

Moreover, variations in terms of housing structure, landuse, land value, demographic composition and infrastructural facilities have been pointed out with the help of this study. landuse planning in terms of distribution, growth, density and composition of population can be done on a micro-level from this model, so that the haphazard growth of suburbs can be minimised by proper suburban planning.

THREE

METHODOLOGY AND DATA ASPECTS

The study has been formulated to explain the residential pattern of suburbs with the help of various quantitative and qualitative techniques in the following manner :

(i) The growth and development of suburbs has been analysed through the process of migration;
(ii) The morphological growth of suburbs has been examined in terms of change in landuse and land value;
(iii) The change in demographic aspects includes an examination of distribution, density, growth rate and composition of population;
(iv) The available infrastructure facilities, like metalled road, frequency of city bus service, availability of shops and other amenities, have been analysed;
(v) Socio-economic status of residents of core villages and housing societies has been described;
(vi) Behaviour and life-style, interaction pattern, perception of the residents of core villages and housing societies have been compared;
(vii) Housing characteristics of core villages and housing societies have been examined;
(viii) A 'Socio-Demographic' model has been proposed for the understanding of a composite spatial organisation of residents in terms of socio-economic status, behaviour and life-style and housing characteristics of the suburbs of Ahmedabad.

Various tables, both univariate and bivariate type simple and composite graphs, several types of maps of distribution, flow map, three dimensional isometric map, statistical tests (chi-square and t-test), correlation analysis, analysis of variance have been used to test the hypotheses.

1. Methodology regarding the demarcation of the suburbs and the study area:

(a) Suburbanisation Index; (b) Residential Index

The demarcation of the outer limit of the suburbs is very essential to classify types of suburbs. In India the concept of suburbs has not been precisely made clear till 1971. In the census of 1971 a new concept of urban unit has been developed, viz. 'Standard Urban Area'. The 'Standard Urban Area' may be defined as the projected growth-area of a city or a town (as it would be in 1991), taking into account of the towns, villages which will be merged into it and also the intervening areas which will also be included into it.

According to the census of 1971, 'Standard Urban Areas' include all areas which have any of the following characteristics:

(i) Predominantly urban landuse;
(ii) Intensive interaction with the urban centres as reflected in communication for the purpose of work and secondary education, extension of city bus service, sale of commodities like milk dairy products, vegetables, and purchase of food grains, cloth and general provision etc. by the consumers directly;
(iii) Anticipated urban growth as a result of locational decisions relating to industry, market, transport and communication;
(iv) Existing big villages with a large proportion of working force engaged in non-agricultural categories.

According to the Census of 1971, Ahmedabad Standard Urban Area has sixtysix components or maujas (revenue boundaries) of which fourteen maujas are towns and fiftytwo maujas are villages (Appendix 1). Total area of 'Standard Urban Area' is nearly 400 square kilometer and whole area is within a range of fifteen kilometers from Ahmedabad city.

Fig. 3.3

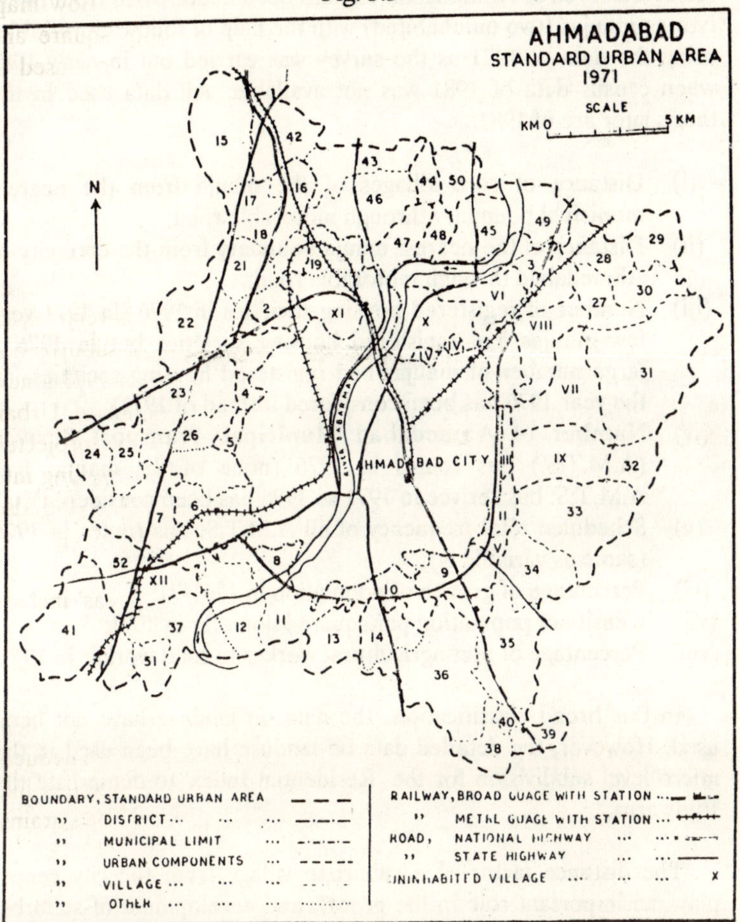

SOURCE : CENSUS OF INDIA, 1971.

(a) *Demarcation and classification of suburbs*

This study focusses on the suburbs of the new residential area of Ahmedabad, situated in the western side of the river Sabarmati. There are thirtyone maujas in the west of Sabarmati (excluding Ahmedabad city) of which eight are in Gandhinagar district. So to consider the maximum limit of suburbs of Ahmedabad city, eight maujas of Gandhinagar district are eliminated as they are under the influence of Gandhinagar. Therefore, a classification of suburbs of residential area of Ahmedabad city has been demarcated from twentythree maujas (two uninhabited) with the help of following variables using the data of 1971 as the survey was carried out in early 1982 when census data of 1981 was not available. All data used in the thesis later are of 1981.

(i) Distance of core villages of the mauja from the nearest municipal boundary through motorable road;
(ii) Distance of the nearest mauja boundary from the core city of Ahmedabad through motorable road;
(iii) Number of registered housing societies in 1976 (in 1971 very few maujas had registered housing societies but in 1976 a large number of maujas had registered housing societies, so the year 1976 has been considered instead of 1971);
(iv) Number of Ahmedabad Municipal Transport Serivce (A.M.T.S.) bus routes in 1976 (none of the maujas had A.M.T.S. bus serivce in 1971 so 1976 has been considered);
(v) Scheduled total frequency of all A.M.T.S. bus routes in 1976 (same as variable iv);
(vi) Percentage of growth of population in 1961-71;
(vii) Density of population per square kilometer in 1971;
(viii) Percentage of non-agricultural worker to total worker in 1971.

In this broad classification, the data on landuse have not been used. However, the detailed data on landuse have been used in the micro-level subdivision for the 'Residential Index' to demarcate the study area.

The distance factor of a suburban village from the city centre plays an important role in the growth and development of suburbs. Variables one and two, i.e. distance of the core village from the municipal boundary and distance of the nearest mauja boundary from the core city deal with the same distance factor but they have somewhat different role of play in the growth of a mauja. The distance of the core village from the municipal boundary is the controlling factor of the transfer of agricultural land into any type of urban use, e.g. residential, industrial or commercial. The villagers sell their agricultural land at a faster rate when they are situated near the boundary of the city and have more interaction with the city. On the other hand, nearness of the mauja boundary from the core city helps in the development of mainly residential area and of other urban uses in suburbs. Initially, areas near to the city limit and core-city develop as the residential area of suburbs, where residents enjoy facilities both of the city and of suburbs. Areas nearer to the city centre are preferred by the new suburban residents because the centre is the place of work, education and shopping.

The number of registered housing societies is one of the important variables for demarcation of the suburb, because this study has focused on the residential pattern. It is always possible that, all registered housing societies may not come into existence because some of them will not be built for speculative purposes. This aspect will be discussed in Chapter VI. Registration of housing societies in the suburb gives some idea about demand for urban land.

Easy accessibility is one of the important factors for demarcation of suburbs. This indicates the availability of transport facilities. In Ahmedabad, 'Ahmedabad Municipal Transport Service' (AMTS) has their bus routes in areas outside the municipal limit. This is the only cheap public transport system which connects suburbs with various parts of the city. Therefore, variables (iv) and (v), i.e. number of A.M.T.S. bus routes and scheduled frequency of bus routes give an idea of the accessibility of a mauja in relation to other parts of the city.

Variables (vi), (vii) and (viii) i.e. percentage of growth of population, density of population and percentage of non-agricultural worker deal with demographic characteristics. High growth and density of population indicate the migration of population in the suburban mauja while higher percentage of these variables indicates more urbanisation.

An index, i.e. 'Suburbanisation Index' (S.I) has been drawn to demarcate the suburbs, using all these variables. The following steps have been used to formulate the 'Suburbanisation Index'.

Firstly, mean and standard deviation of each variable have been calculated and average value of any variable has been 10 points and +1 point for half of a standard deviation. Points are calculated in a reverse manner for variables related to distance, i.e. longer the distance of a village from the core city or city limit, lower is the point. The maximum point that can be obtained here is 20 and the minimum point is 0 (Appendix 2).

Secondly, each variable has been given weightage according to their importance in relation to demarcation (Appendix II). Thirdly, point of each variable has been multiplied by the value of the weightage of that variable and the obtained score of all variables has been added to get the total score for each mauja. Lastly, this score of respective mauja has been divided by maximum score (720 in this case) and the quotient 'Suburbanisation Index' thus obtained, has been multiplied by 100.

Mean and standard deviation of scores of 'Suburbanisation Index' have been calculated to demarcate variations within the suburb (Appendix II). By using graphically the mean and the standard deviation and the distance from the city limit, four main subdivisions have been made based on the concept of distance-decay (Figure 3.4).

1) Urban suburbs.

2) Moderately urban suburbs.

3) Moderately rural suburbs.

4) Rural suburbs.

The urban suburb is extended nearly one kilometer from the city limit. This type of suburb has very high score of 'Suburbanisation Index' due to its high rate of urbanisation. The 'Suburbanisation Index' score decreases sharply towards the moderately urban suburb which is extended nearly two kilometer towards the rural area from the outer limit of the urban suburb. Though the moderately urban suburb is less urbanised but it has high score in 'Suburbanisation Index' compared to the score in 'Suburbanisation Index' of moderately rural and rural suburbs. These two types of suburbs are very far from the city limit; so they are less urbanised and have low score in 'Suburbanisation Index'.

Fig 3.4

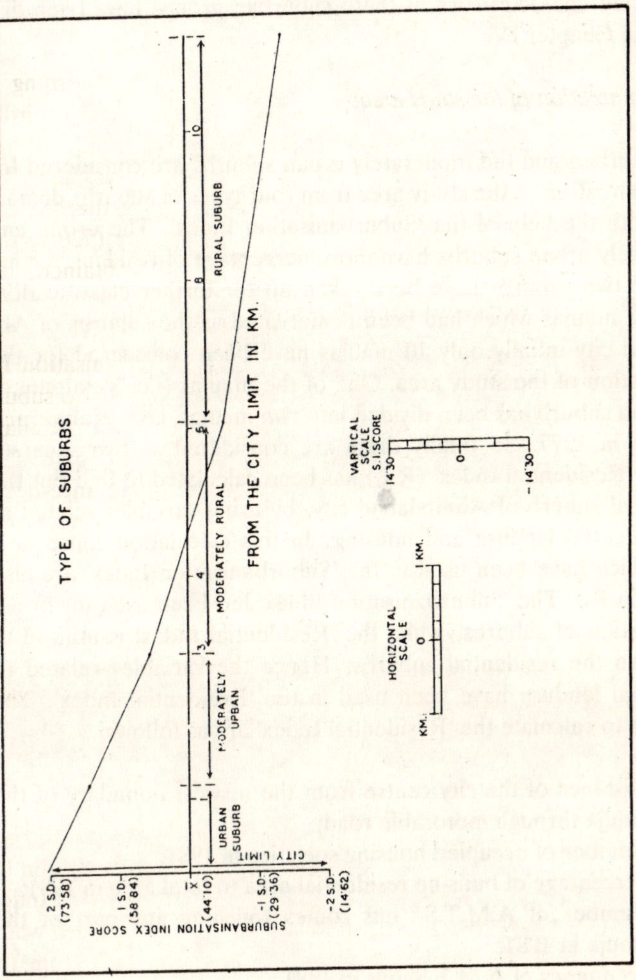

SOURCES: VILLAGE PANCHAYAT, A.M.T.S., DISTRICT CO-OPERATIVE SOCIETY, DISTRICT CENSUS HANDBOOK, 1961, 1971. (APPENDIX - 2)

So it can be stated that score in 'Suburbanisation Index' decreases sharply from the city limit towards the moderately urban suburb and then it declines gradually towards the rural suburbs. The nature and characteristics of these suburban groups have been discussed in Chapter IV.

(b). *Demarcation of the study area*:

The urban and the moderately urban suburbs are considered for the demarcation of the study area from four types of suburbs demarcated with the help of the 'Suburbanisation Index'. The urban and moderately urban suburbs have more percentage of residential land, so these two suburbs have been taken up for further classification. From 21 maujas which had been demarcated as the suburbs of Ahmedabad city intially only 10 maujas have been considered for the demarcation of the study area. One of the maujas (i.e. Vejalpur) of the urban suburb has been divided into two maujas, i.e. Vejalpur and Jodhpur in 1977. So finally these are considered as two separate maujas. 'Residential Index' (R.I.) has been calculated to find out the residential suburb of Ahmedabad city, by using variables related to the residential landuse and housing. In this calculation some variables which have been used in the 'Suburbanisation Index' are also utilised in R.I. The 'Suburbanisation Index' has been used for broad classification of suburbs, while the 'Residential Index' is utilised to categorise the residential suburbs. Hence the variables related to residential landuse have been used in the 'Residential Index'. The variables to calculate the 'Residential Index' are as follows:

(i) Distance of the city centre from the nearest boundary of the mauja through motorable road;
(ii) Number of occupied housing societies in 1981;
(iii) Percentage of built-up residential area to total area in 1981;
(iv) Number of A.M.T.S. bus routes touching any part of the mauja in 1981;
(v) Frequency of A.M.T.S. bus in 1981.

Variables (i), (iv) and (v) i.e. distance from the city centre, number of routes and frequency of A.M.T.S. bus services are common variables for S.I. and R.I. classification. The number of occupied housing societies along with the built up residential area indicates the growth and development of the residential suburbs. As stated earlier, the distance from the city centre is an important factor that controls the growth and development of residential suburbs, because, people prefer to live nearer to the city centre as they commute for work, shopping and other reasons. Accessibility is another important factor. So the number of bus routes and the frequency of buses are again considered to demarcate various types of residential suburbs. The 'Residential Index' has been calculated by following the same method as has been adopted for the calculation of 'Suburbanisation Index'. Only in this case the average value has been given 5 points and + 1 point for 1/2 standard deviation and maximum point is 10 and minimum point is 1 (Appendix 3). The urban and moderately urban suburbs have been divided into following categories (Figure 3.5) :

1) Purely residential
2) Semi-residential
3) Non-residential

Purely residential suburb is more or less extended along the city limit. It has very little horizontal extension but the score in 'Residential Index' is very high. It is mainly due to the growth of areas near the city limit as residential area in the initial stage. The score in the 'Residential Index' declines very sharply from purely residential suburb to semi-residential suburb. It is mainly due to the presence of few housing societies and residential area in this type of suburb. This type of suburb is horizontally extended only upto half a kilometre. Non-residential suburb with very low score in 'Residential Index; covers a very large area of the urban and moderately urban suburb. So, it can be said that purely Residential and Semi-Residential suburbs are extended horizontally less than one kilometre from the city limit.

2. Methods of depicting the residential pattern of the study area with the help of a suggested 'Socio-Demographic' model:

The study area, which has been selected with the help of the 'Residential Index' has been further classified on the basis of morphology, socio-economic status and behaviour of residents. The residential pattern of each mauja has been graphically measured by a three dimensional axis - (a) socio-economic status of residents (x-axis) (b) residents' behaviour and life-style (Y-axis) and (c) housing characteristics (Z-axis). Each axis has been measured in terms of 'Modernity' or 'Traditionality'. The word 'Modern' used in this study, is to indicate residents belief in modern outl ., behaviour and life-style which are different from the age-old traditional belief. The details of these variables related to modernity or traditionality have been discussed in Chapter VII and VIII. These three major factors are based on 53 variables (Appendix 4). Data of these 53 variables have been collected through field survey. Qualitative data have been measured and represented by numerical scores as far as possible to measure the factor of modernity. Since much of the data is behavioural in nature, no attempt has been made to use cluster analysis or discriminant analysis in this classification.

The details of calculation of 'Modernity Index' are as follows :

(i) Frequency of each sub-variable has been calculated separately for each mauja.

(ii) To make it comparable, percentage of each sub-variable, in relation to total of that variable in a mauja, has been calculated (Appendix 4).

(iii) Weightage has been allotted to each sub-variable (Appendix V). In case of some variables, 'Principle of Scarcity' has been used, i.e. the higher is the scarcity of a variable , the higher is the weightage assigned to that variable. In all variables scores have been given on an arbitrary relative scale.

(iv) Weightage of sub-variables have been multiplied with their percentage, the total of which indicates the points for that variables. The available points of all variabales, forming each factor, have been added to get the total points for that factor. Total points of the three major factors have been calculated for all selected maujas and also for the study area as a whole.

(v) The total points of each factor in each mauja have been divided by the total points of each factor of the study area, to get the 'Modernity Index'. The obtained quotient has been multiplied by 100. In the scale of 'Modernity Index', if the value of a factor in a mauja is more than 100, then the factor is called 'Modern' in that mauja and if it is less than 100, then it has been termed as 'Traditional' (Appendix 6).

Fig 3.5

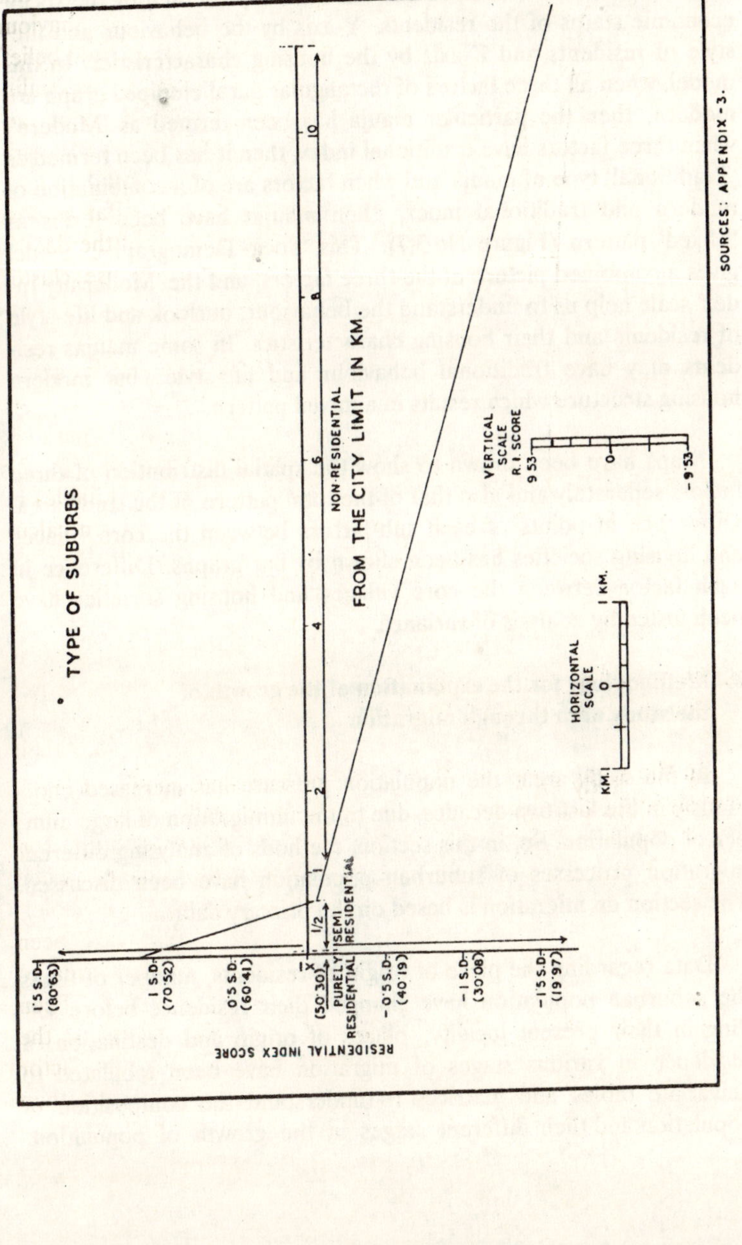
DIVISION AND DISTRIBUTION OF URBAN AND MODERATELY URBAN SUBURBS
TYPE OF SUBURBS
FROM THE CITY LIMIT IN KM.
NON-RESIDENTIAL
SEMI RESIDENTIAL
PURELY RESIDENTIAL
RESIDENTIAL INDEX SCORE
1·5 S.D. (80·63)
1 S.D. (70·52)
0·5 S.D. (60·41)
X̄ (50·30)
-0·5 S.D. (40·19)
-1 S.D. (30·08)
-1·5 S.D. (19·97)
VERTICAL SCALE R.I. SCORE
HORIZONTAL SCALE
KM. 1 0 1 KM.
9·53 0 -9·53
SOURCES: APPENDIX - 3.

The 'Socio-Demographic' model has been represented in a rectangular parallelopiped graph, showing the three factors in three dimensions (Figure No.3.6), in each mauja of the study area. In this three dimensional model, X-axis has been represented by the socio-economic status of the residents, Y-axis by the behaviour and life-style of residents and Z-axis by the housing characteristics. In this model, when all three factors of rectangular parallelopiped graph are modern, then the particular mauja has been termed as 'Modern', when three factors have traditional index, then it has been termed as 'Traditional' type of mauja and when factors are of a combination of modern and traditional index, then maujas have been terms as 'Mixed' pattern (Figure No.3.7). This 'Socio-Demographic' model gives a combined picture of the three factors, and the 'Modernity Index' scale help us to understand the behaviour, ourlook and life-style of residents and their housing characteristics. In some maujas residents may have traditional behaviour and life-style but modern housing structure which results in a mixed pattern.

Maps have been drawn to show the spatial distribution of three factors separately and also that of the total pattern of the study area. Difference in points of each sub-factors between the core villages and housing societies has been shown by bar graphs. Difference in each factor between the core villages and housing societies have been tested by analysis of variance.

3. Methodology for the explanation of the growth of the study area through migration

In the study area, the population pressure has increased enormously in the last two decades, due to the immigration of large number of population. So, in this section, methods of analysing different migration processes of suburban population have been discussed. The section on migration is based on the primary data.

Data regarding the place of origin of residents, number of times the suburban population have changed their residence before settling in their present locality, places of origin and destination of residence in various stages of migration have been tabulated to univariate tables and matrices to understand the composition of population and their different stages in the growth of population.

Fig 3.6

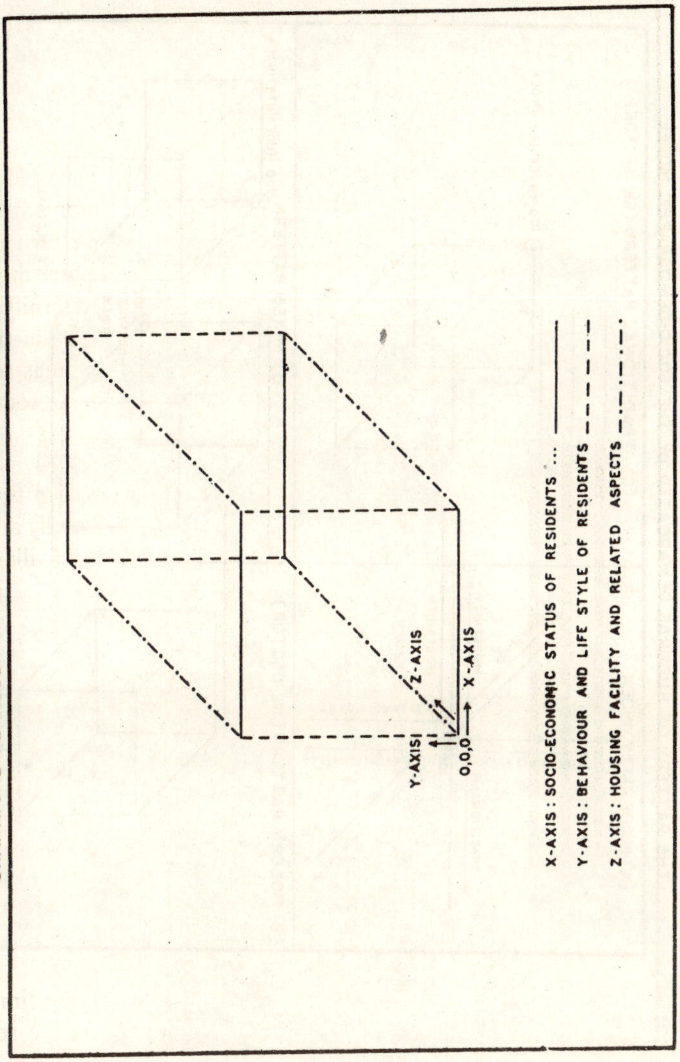
GRAPHICAL REPRESENTATION OF THE SOCIO-DEMOGRAPHIC MODEL
Y-AXIS
Z-AXIS
0,0,0
X-AXIS
X-AXIS : SOCIO-ECONOMIC STATUS OF RESIDENTS ...
Y-AXIS : BEHAVIOUR AND LIFE STYLE OF RESIDENTS
Z-AXIS : HOUSING FACILITY AND RELATED ASPECTS

Fig. 3.7

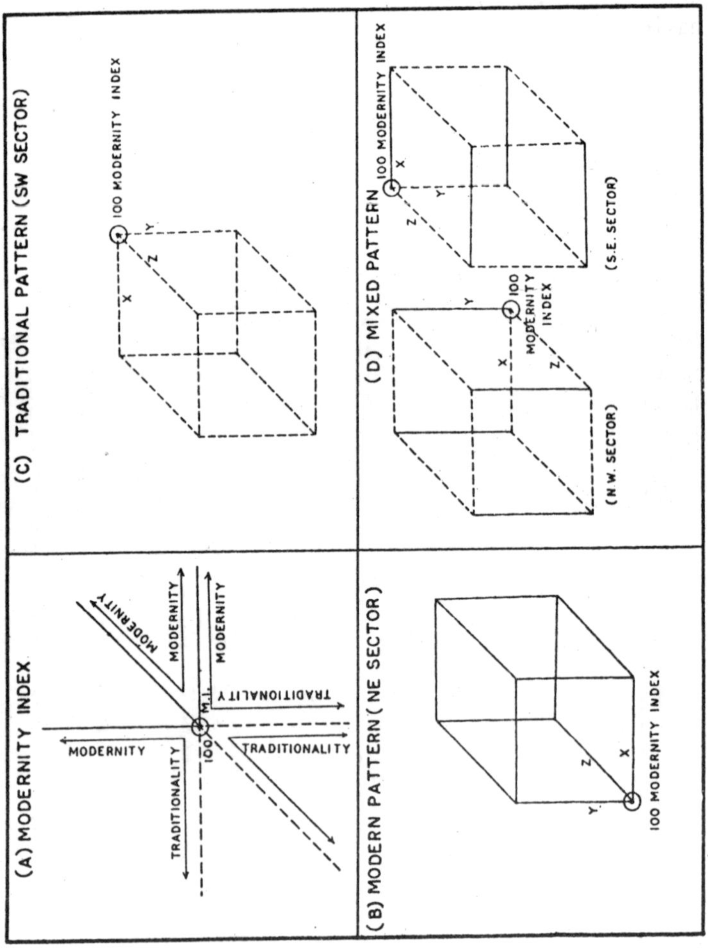
THE PATTERN OF SUBURBS DEMARCATED BY THE SOCIO-DEMOGRAPHIC MODEL
(A) MODERNITY INDEX
MODERNITY
TRADITIONALITY
MODERNITY
MODERNITY
MODERNITY
TRADITIONALITY
TRADITIONALITY
M.I.
100
(B) MODERN PATTERN (NE SECTOR)
100 MODERNITY INDEX
X
Y
Z
(C) TRADITIONAL PATTERN (SW SECTOR)
100 MODERNITY INDEX
X
Y
Z
(D) MIXED PATTERN
100 MODERNITY INDEX
100 MODERNITY INDEX
(N.W. SECTOR)
(S.E. SECTOR)

One flow chart has been drawn to understand the migration of the residents of the study area in different stages. A generalised model has been formulated to show the distribution of population on space, in their place of origin, of different stages of migration. This model has been drawn with the help of dot method.

4. **Methodology regarding the analysis of morphological and demographic growth**

This section of the study is based mostly on the secondary data of landuse, demography and infrastructure of the area. The distribution of various types of landuse has been analysed over a period of time (1961-81). This aspect of landuse has been explained with the help of choropleth maps, univariate and bivariate tables, graphs, etc. Isopleth maps of landuse over time (1971, 1981) and space have been drawn to understand the relationship between the change in landuse and land value. An isometric map of land value has also been drawn to analyse the land value per 0.25 square kilometre area, and to bring out clearly the three dimensional effect of the land value proportionate to height.

A super imposed sphere map over time (1961, 1981) and space has been drawn to explain the distribution of population.

The density of population has been analysed in two ways, e.g. gross density and net density. Gross density is the ratio between total population and total area while net density is the ratio between residential population and residential area. Both these densities have been examined with the help of mapping over time and space. The distribution of maps has been used to understand the growth rate of population for 1961-81. Literacy, sex ratio and work participation have been studied over space and time to know the quality of population. The distribution of non-agricultural workers has been analysed by a series of maps over time (1961, 1981).

The distribution of amenities of the study area has been looked into with the help of several maps. An isopleth map has been measured in relation to the alignment of metalled roads. Following the method of L.D. Stamp, distance factor has been considered in preparing this map. So a distance of 0.25 km or 250 metres from metalled road has been demarcated as a very highly accessible zone, considering that people can cover this distance to catch public transport without any difficulty in the study area. In this scale of 0.25 km four other zones have been demarcated in the following manner:

(i) Very high accessibility-area upto 250 metres from the metalled road;
(ii) High accessibility - area between 251 to 500 metres;
(iii) Average accessibility - area between 501 to 750 metres;
(iv) Low accessibility - area between 751 metres to 1 km;
(v) Very low accessibility - area more than 1 km from the metalled road.

Connectivity of the study area has been further analysed with the help of frequency of city buses. One flow map has been drawn to show the frequency of buses. Another connectivity map has been drawn to show bus routes. In this map the zone of influence of bus routes has been projected upto half a kilometre (highly accessible zone) in all sides of it. Considering, residents residing at that distance can take the advantage of bus routes when required.

Data on other amenities like education, medical, religious facilities, recreation, shopping facilities (number of authorised shops), post and telegraph, bank have been collected from different sources. Weightage has been assigned to the type of amenities to scarcity method, i.e. higher is the weightage, lower is its availability (Appendix 7). Weighted points for amenities have been multiplied by the frequency of each amenity and then added to get total points. Finally, a choropleth map has been drawn to show the spatial distribution of amenities.

All these choropleth maps have been prepared in a scale of five point index. The average class was defined to include values from mean minus one half of the standard deviation to mean plus one half of the standard deviation. Thus range of this average class equals to one standard deviation. The next classes, i.e. higher and lower called high class, i.e. one standard deviation. The extreme classes of very high and of very low have been kept without limits at one end. In some maps, where range is large by above method there the range has been made one of the standard deviation. In such cases also five classes are made.

Simple correlation analysis and chi-square testing have been made to understand whether morphological aspects, e.g. growth of residential area, non-agricultural worker, bus frequency, amenities have any relationship with each other.

5. Methodology regarding perception attachment and interaction pattern of the suburban residents

This part of the study is solely based on the primary data collected from field work. The interaction pattern of residents has been measured under two heads, i.e. shopping and frequently visited place. These two heads have been used to find out the difference in interaction pattern of the residents of core villages and housing societies.

The general physical and behavioural quality of each household and its surrounding area has been given score on a five point scale, both by residents and by the investigator (author in this case) based on predetermined qualitative norms (Appendix 8). The total assessment of both physical and behavioural aspects has been defined as 'Perception Index of Residents' or 'Investigator' (PIR and PII) and is based on variables like type of locality, shopping facility, amenities and behaviour aspects of residents. The scores of perception indices have been compared by t-test to find out whether there is really any difference of opinion regarding the total environmental quality between the residents and the investigator of maujas. Another t-test of perception index has been performed to compare the opinion between residents of core villages and of new suburban areas.

Various uni-variate and bi-variate tables have been made to analyse images of residents regarding different parts of Ahmedabad city and the study area. Tables also have been made to analyse reasons of likings and dislikings of neighbourhoods by residents.

A number of similar variables has been agglomerated to form an 'Attachment Index'. This index is based on attachment of residents to their own locality, variables like place of birth, duration in present locality, relationship with neighbours, preference to own area, considering own locality as neighbourhood, interaction within own locality in terms of work, education, visiting friends and relatives. With the help of t-test, and 'Attachment Index' of residents of core villages and those of new suburban area has been compared.

Most of the primary data have been displayed on maps of maujas of the study area, to get an idea of the distribution of variables over space which is important for planning.

57 Methodology and Data Aspects

6. Data Aspects

a. *Type and sources*

The study is based on both primary and secondary data. Primary data have been collected through field survey which has been done by the author from the month of July 1982 to January 1983. A field questionnaire has been formulated, covering following aspects to get the primary data. The basic information is as follows:

(i) General information regarding the head of the family, e.g. place of birth, age, sex, caste, level of education, occupation, monthly income, religion, mother tongue, food habit, etc.
(ii) General information regarding the family, e.g. total members, type of family, level of education, occupation, monthly income of each family member and also their relation with the head of the family;
(iii) Behaviour and life-style of the family members, e.g. reading, recreation and savings habit, behaviour of raising children, use of consumer goods and possession of modern goods;
(iv) Housing characteristics, e.g. available facilities in the house, number of rooms, built-up area, rent, selling value, approach road, water supply, electricity, etc.
(v) Interaction pattern of residents, e.g. place of shopping, work, education, recreation, areas visited to meet friends, relatives, frequently visited places for social purpose;
(vi) Perception of residents regarding their own area, other areas of suburbs and other parts of Ahmedabad city, likes and dislikes of present locality, cause of migration from previous

area, image of city and suburbs, etc.

(vii) Perception of investigator in terms of locality, housing condition, behaviour of people, shopping facilities and other amenities.

The primary data are qualitative in nature because these data measure behaviour, life-style, perception of residents of the study area.

The secondary data on landuse was sufficiently available by plot over time. The recent data (last ten years i.e. 1971-81) are available from the panchayat office of each mauja, while old records are available from Taluka Development Office.

All information regarding land are recorded in the revenue register of village panchayat known as Village Form (VF). V.F. 7-12 have records regarding ownership and use of land. These data have been used to draw the landuse maps and to give a clear idea of changes of landuse over a period of time. In this study plot-wise landuse data of a eight maujas are collected for the years 1961, 1971, 1981 to have an idea of changing agricultural land into residential one in two decades (1961-81).

The demand register (V.F.18) of village Panchayat office records the number of residential houses. It records the name of each housing society It also keeps records of number of houses in the village core.

Data on land value are very important to find out the pressure of population on land and the choice of a particular mauja or locality by people. Data regarding land value are available in V.F.6 which registers conversion process of agricultural land for non-agricultural uses. V.F.6 records in detail the area of transaction, amount paid and name of the buyer and the seller. Another source of getting the data of land value in the study area is through interviewing various estate agents in the market.

Data regarding different aspects of population have been provided by the census of India. Census provides data regarding total population, literate persons, sex, total workers and workers engaged in different sectors. These data are collected for 1961, 1971, 1981. These data of population help us to analyse the distribution, density, growth and composition of population.

Secondary data regarding amenities have been collected from several sources. Data regarding educational institutes, medical centres, religious places, post and telegraph offices, banks, etc. are available from Ahmedabad Urban Development Authority Office.

Data regarding transport facilities like number of city bus routes and their scheduled frequency are available from Ahmedabad Municipal Transport Service (A.M.T.S.) office.

(b) *Problems of data*

Most of the published data are available in an aggregated level which do not cater to the need of this study. Some data on perception which are essential for this study are not available. As a result, the need of primary data is felt very acutely in the study.

Data regarding any type of institutional quarters, i.e. railway quarter, industrial quarter and unauthorised slums are not available in the Demand Register of village panchayat office.

Data regarding landvalue are not very reliable. Details of registered transaction are not accurate in terms of price and date. The actual price vary from the registered due to undeclared premium. The real date of transaction vary between the actual date of registration and the date of 'banakhat' (an agreement between buyers and sellers) which gives details of terms and conditions including the land value. In the suburban land market of Ahmedabad, normally 'banakhat' is done before actual transaction gets registered. 'Banakhat' remains valid for 5 years. Another serious limitation of data is that there is no way to trace speculative transaction. Estate agents can give a better idea of the actual price of an area. But the main problem of land value available from estate agents is that it is not consistent because different estate agents give different values of the same area. So an average land value has to be taken into consideration.

Regarding the data on workers in the census, the problem is that in each decade the definition has been changed and it is practically impossible to compare data of one decade with the other. These census data do not give any idea of population living in various residential areas, i.e. core villages and new suburban area. There is no data regarding the place of birth of people, caste, religion, mother tongue, age, level of education and occupation, etc.

In case of amenities, there is no secondary data of number of authorised shops and also number of passenger per day and per bus. There are also problems in collection of the primary data because, in some cases, residents do not respond fully.

7. Sampling Technique

In order to have a close view of the residents of the study area in terms of their socio-economic status, behaviour and life-style and their housing characteristics, primary data is very much needed and that has been collected with the help of field questionnaire. As time, funds and man power are limited, so 1500 sample has been taken purposely. 1500 households (excluding those who are living in quarters of various institutions) living in 1500 residential houses have been interviewed. This is nearly 6% of total houses of the study area in 1982 (24322).

For the purpose of sampling, 1500 houses are proportionately divided into 8 maujas in proportion to their total occupied residential houses. Sample houses of each mauja have been again stratified into proportion to the number of houses in the core village, housing society and other residential type like colonies, authorised slums etc. Total samples of all housing societies in a mauja have been again stratified in proportion to the number of houses in each society. In core villages samples have been selected on stratified random method to get representative sample from each caste. In housing societies and other residential structures, samples have been selected on random method.

8. Development of Hypotheses

The following hypotheses have been developed in this study and these have been tested either qualitatively or quantitatively to arrive at a conclusion.

(i) Most of the residents of the study area have migrated in their present location through the process of step-wise migration.
(ii) A large amount of agricultural land has been changed into non-agricultural land, mainly into residential land during 1961-81 in the study area and the growth of residential landuse is not evenly distributed.
(iii) The population distribution in terms of the area, is concentrated in nature both in 1961 and 1981.
(iv) Growth of population is positively related to the growth of residential and to non-agricultural area.
(v) Frequency of buses, distribution of shops and other amenities are equally distributed.
(vi) There is difference between the residents of core villages and housing societies regarding their socio-economic status, behaviour and life-style and housing characteristics.
(vii) The residential pattern in suburbs is of mix-type with modern housing structure and with residents having traditional behaviour.

The experience from the field survey has been very helpful to understand the non-measurable behavioural factor of people which is responsible for residential pattern of the study area.

FOUR

AHMEDABAD CITY, ITS SUBURBS AND THE STUDY AREA

This study will focus on the newly developed residential suburbs of Ahmedabad city. So the regional location, the growth and development and the characteristics of Ahmedabad city and its suburbs are discussed briefly in following lines.

1. The regional location of Ahmedabad and the historical perspective of development of Ahmedabad

Ahmedabad is the largest city of Gujarat State and the sixth in whole India. Population of Ahmedabad city is 2,123,831 in 1981. Ahmedabad is situated in western part of India (Figure 4.8) and it is connected with various parts of the country by railway lines (both by broad gauge and by meter gauge) and roads. Though the state capital has been shifted to Gandhinagar in 1971, which is situated 30 km north-east of Ahmedabad, the city is the main centre of education, professional services, industry, trade and commerce in the State.

Ahmedabad is situated on the plain of the river Sabarmati. The river Sabarmati divides Ahmedabad into two distinct parts. Though the contour of Ahmedabad is more or less plain, but there are some hillocks in the western part of the river Sabarmati. The river Sabarmati acted as a natural barrier in the growth of the city towards west in the past. The original core-city of Ahmedabad is located in the eastern side of the river Sabarmati, which has some characteristics of pre-industrial cities with residential-cum-commercial area and this core-city is merged on the eastern side with its modern industrial zone with textile as the main industry. The western side of the river Sabarmati has started to develop after 1930 or so and that is truly residential in character, except in the extreme north where some industries are located (Figure 4.9).

Fig. 4.8

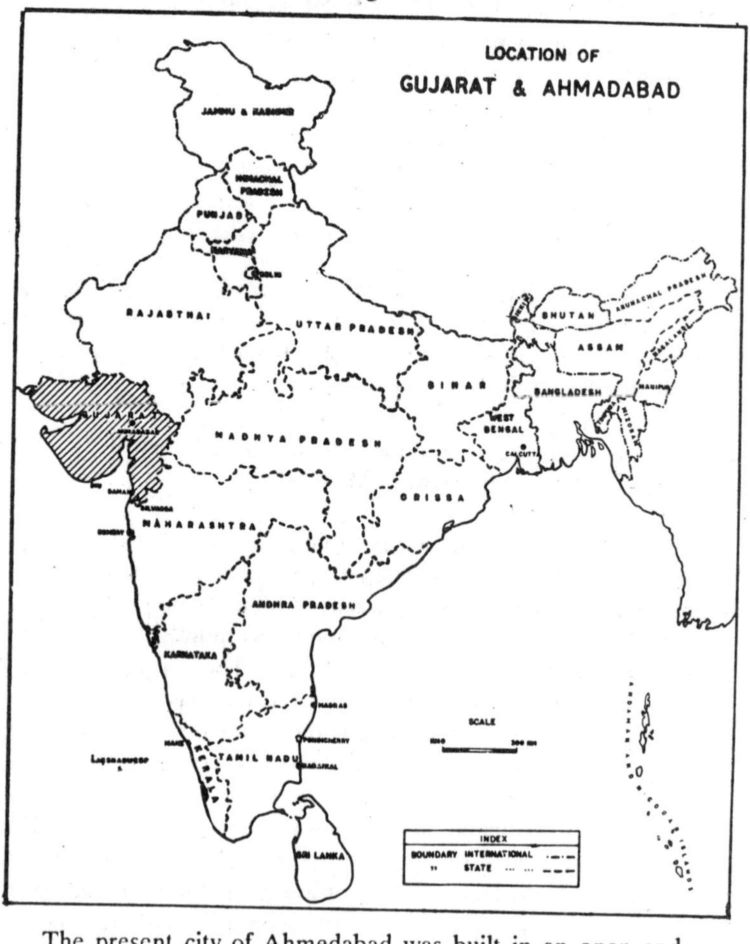

The present city of Ahmedabad was built in an open and spacious plain to the east of the river Sabarmati in 1411 A.D. In 1847 the city was enclosed by a fort wall of six miles in circumference

Fig. 4.9

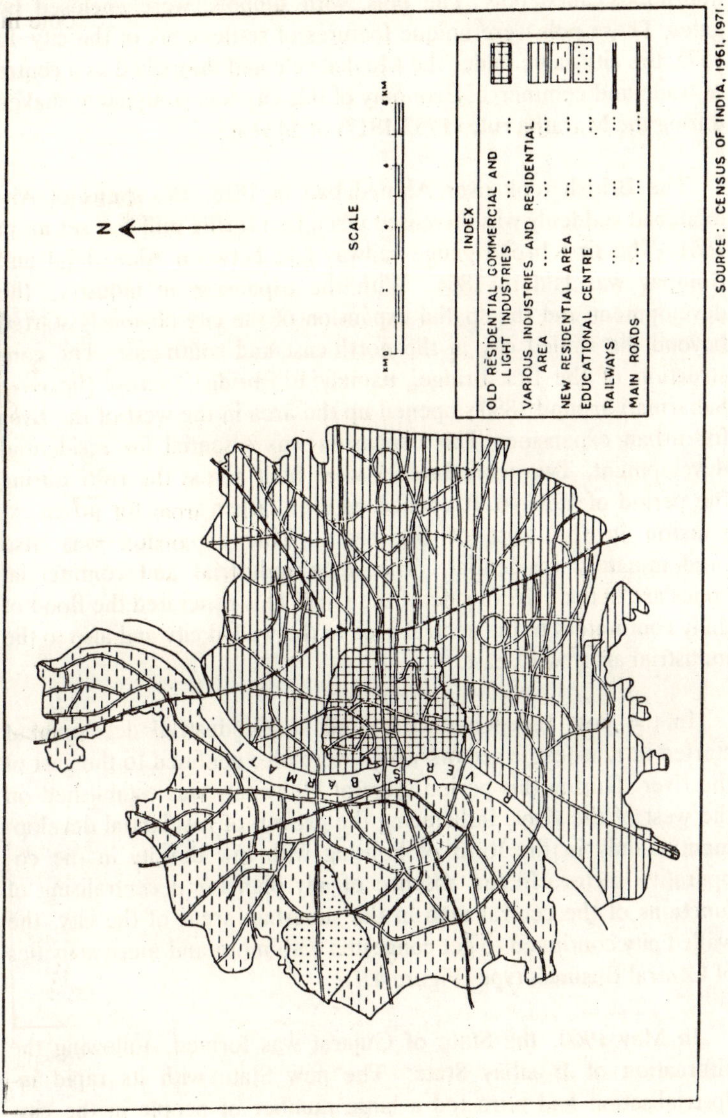
BROAD LAND USE MAP OF AHMADABAD 1981
N
SCALE
INDEX
OLD RESIDENTIAL COMMERCIAL AND LIGHT INDUSTRIES ...
VARIOUS INDUSTRIES AND RESIDENTIAL AREA ...
NEW RESIDENTIAL AREA ...
EDUCATIONAL CENTRE ...
RAILWAYS ...
MAIN ROADS ...
RIVER SABARMATI
SOURCE : CENSUS OF INDIA, 1961, 1971.

having twelve gates. The city then had a fort, 17 Chaklas (Central squares where four road meet) and about 80 principal 'bazars' or markets. The residential quarters which grew around the fort were segregated by castes, religion and occupation into number of neighbourhoods called 'pols'. The pols, with subpols, were enclosed by gates. These pols were unique features of settlements of the city. In 1572, the city came under the Mughal rule and flourished as a centre of trade and commerce. Economy of this city was completely shaken during the Maratha rule (1753-1817) of 54 years.

The British took over Ahmedabad in 1818. the status of Ahmedabad suddenly was increased when first textile mill was set up in 1861. The first broad gauge railway line between Ahmedabd and Bombay was laid in 1864. With the expansion in industry, the development and the spatial expansion of the city obviously started beyond the walled city in the north-east and south-east. The construction of the first bridge, namely Ellisbridge, across the river Sabarmati around 1880's opened up the area in the west of the river for urban expansion. This expansion was essential for residential development. Two more bridges were built across the river during the period of 1931-46. It opened up even larger areas for urban expansion in the west of the river. This expansion was also predominantly residential. The main industrial and commercial zones are in the eastern side of the river. This generated the flood of daily commuters to the central area in the walled city and also to the industrial area on its east.

In the post Independence period, the industrial development started in non-textile sectors which were also confined to the east of the river. During this period Gujarat University was established on the west of the river, influencing the growth of residential development mainly in the form of housing with low density in the cooperative sector. In the absence of any plan for decentralising of functions of the central area in the suburban areas of the city, the walled city continued to get congested with more and more activities of Central Business type (Figure 4.9).

In May 1960, the State of Gujarat was formed, following the bifurcation of Bombay State. The new State with its rapid industrialisation had attracted a large number of people in the city.

The residential pattern in the western side has changed and new trend has started in last two decades, that is of high rise development in the form of flats and apartments instead of bungalows and low rise complex. This obviously is the result of increasing population pressure on land, and has raised the population densities in these residential areas.

As the pressure of population started mounting within the city limit. people started to live in areas outside the city limit. New settlements have started developing in the areas which were formerly agricultural land. These areas were originally owned by the residents of core villages. The villagers have started to sell these lands to residents of the city when new suburban development has started already in areas outside the city limit. In the first stage, plots of land nearer to the city limit were sold and gradually most of the plots were sold for suburban development in areas away from the city limit.

The history of development of suburbs in Ahmedabad can be found as early as Muslim period, when the 'pols' of the the cities had a number of suburbs. These suburbs grew around original settlements to house the increasing population of the city. The suburbs, situated outside the city limit, were villages. But this suburban expansion was not confined only to the areas on the eastern side of the river but extended even to the western side of the river. These suburban villages, originally situated outside the city, were gradually engulfed by the city, but the core villages can still be detected in their old structure of houses, composition and behaviour of population. They look like islands in an urban area as 'Urban Villages' (Figure 4.10). Similar process is seen also in the development of suburbs. Settlements start growing near the city limit and eventually spreads around villages and those villages will be included in the city area later on.

In different phases, a large position of the suburbs of Ahmedabad has been engulfed by the city,thereby increasing the area of the city of Ahmedabad (Table 4.1). Ahmedabad city had low density of population till 1941, except in 1881. The density of population has been increased in 1961 and 1981 due to the formation of the Gujarat State. In this century (1901-1981) population of Ahmedabad city has

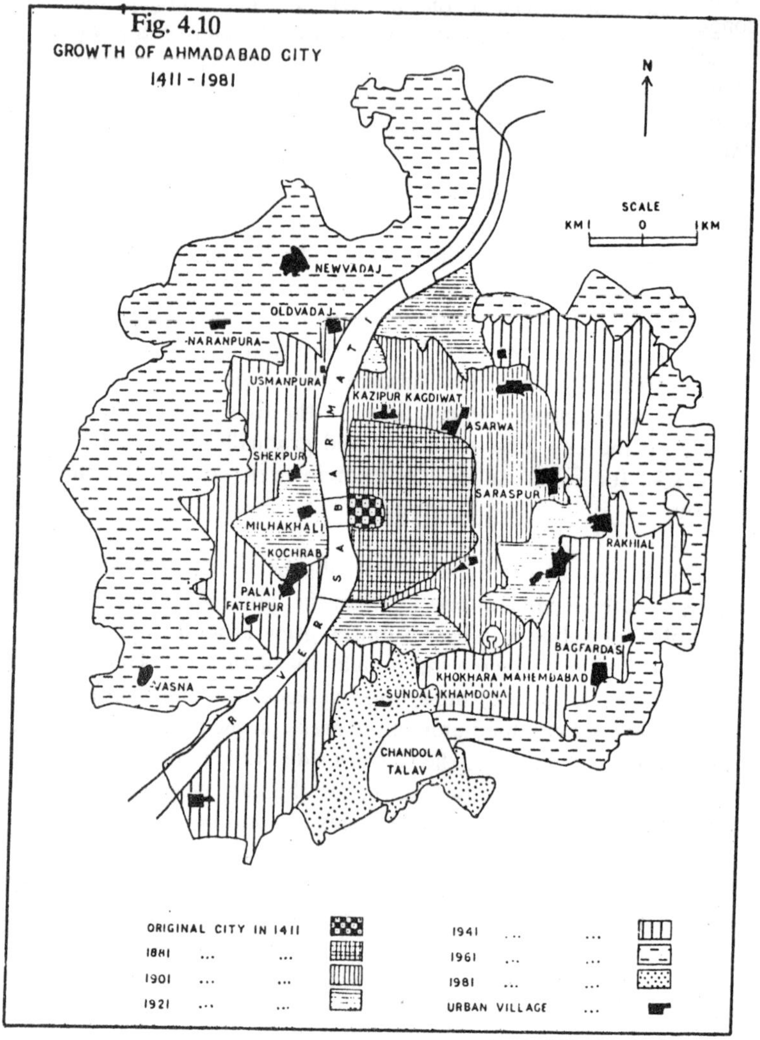

Fig. 4.10
GROWTH OF AHMADABAD CITY
1411 - 1981

SOURCE : DISTRICT CENSUS HANDBOOK, 1961, 1981.

increased 1042% while area has increased 562%. So the growth of the area of Ahmedabad city is slightly lower than its population in the last 100 years. It indicates the pressure of population on the urban land and the growth of suburban residential areas in the recent years. It is interesting to note the density of population of Ahmedabad city has increased only 0.1% in 1981 than as it was in 1881.

So it can be stated that, in 1881, Ahmedabad city was also a very congested city. This congestion has caused the expansion of the city in various stages. But in 1961-81 the city experienced an enormous growth of population in relation to its total area, which is mainly due to the rapid industrialisation of the city.

TABLE 4.1 : **Area and population of Ahmedabad City 1411-1981**

Year	Area in Sq.KM	Growth rate of area	Population	Growth rate of popula-tion	Density per Sq.Km²
1411	0.38		N.A		
1811	5.95	150.9	127,621	45.7	21,449
1901	14.93	60.5	185,889	47.4	12,451
1921	23.96	119.0	274,007	115.8	11,436
1941	52.47	77.2	591,267	94.5	11,269
1961	92.98	6.3	1,149,918	84.7	12,368
1981	98.84		2,123,831		21,466

N.A. : Not Available

Source: Census report 1961, 1981

2. Distribution and the Characteristics of the suburbs of Ahmedabad city

Keeping pace with the process of urbanisation, the suburb, surrounding the city, undergoes changes from a completely rural environment to an urban one. so it is very much essential to find out the characteristics of various types of suburbs to know the stages that villages around a city may pass through.

In the previous chapter (III), the methods of demarcation and distribution of different types of suburb have been discussed. Two indices, i.e. 'Suburbanisation Index' and 'Residential Index' have been formulated to depict the distribution and characteristics of suburbs. This classification has been worked out with reference to the suburbs of the residential areas of Ahmedabad city (Figure 4.9).

(a) *Characteristics of different types of suburbs*

The 'Suburbanisation Index' has been used to divide maujas (revenue boundary of villages) into four zones, i.e. rural suburb, moderately rural suburb, moderately urban suburb, urban suburb (Figure 4.11). Different characteristics of suburbs are described briefly in the following section (Table 4.2).

Fig 4.11

CHARACTERISTICS OF SUBURBS OF AHMADABAD
1971

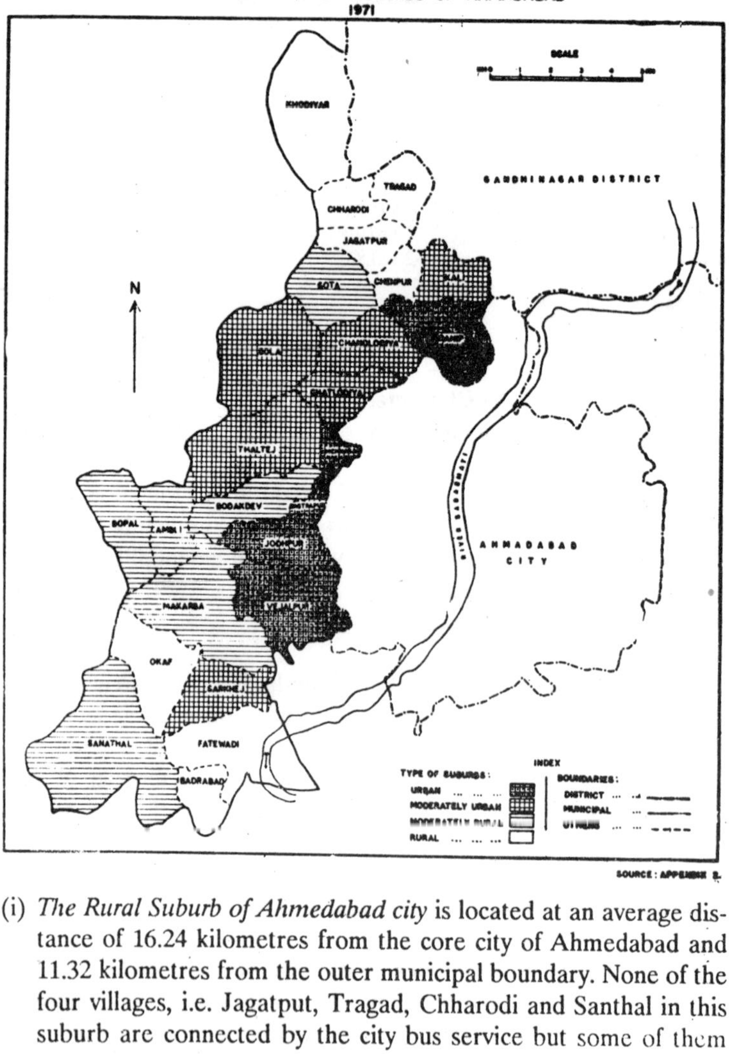

SOURCE : APPENDIX 3.

(i) *The Rural Suburb of Ahmedabad city* is located at an average distance of 16.24 kilometres from the core city of Ahmedabad and 11.32 kilometres from the outer municipal boundary. None of the four villages, i.e. Jagatput, Tragad, Chharodi and Santhal in this suburb are connected by the city bus service but some of them have connection with the city by 'State Transport' bus service. The presence of non-agricultural fallow, which are occupied by unbuilt housing societies, indicates the future possibilities of suburbanisation. The average number of registered housing societies is few, that is less than 1, and none of them is built due to low demand of residential houses in this type of rural suburbs. Though most of the population (63%) are engaged in agricultural activities, there is daily movement from the village to the city to supply perishable goods like milk and vegetables, especially in winter. Growth of population is 49% during 1961-71 and density of population in 1971 is 231 per sq.km. Both the figures are very low in the rural suburbs. As a whole, this suburb is still rural in physical and social characteristics. As the people of this type of suburbs have low social and personal interaction with the city, they are still rural in their behaviour and life-style. The residents do not have urban goods like tape recorder, refrigerator, television or scooter.

(ii) *The Moderately Rural Suburb* is nearer to both the city-centre (11 km) and the city limit (6 km) than the rural suburb. This type of suburb includes seven maujas namely Ambli, Bodakdev, Gota, Makarba, Bhopal, Chenpur and Fatewadi. Of these only three maujas are connected by the city bus service (Appendix 2). The average number of routes (only one) and the frequency (nine) are also very low. In this type of suburbs the farming is changed in the sense that here a part of land is devoted to commercial farming like vegetables and flowers. Though the majority of people (61%) are engaged in agricultural activities, some of them have started seeking subsidiary jobs in the city. Some residents have started small business of their own, like cycle repair shops on the road side, barber's shop, vendors, etc, in the urban suburbs and in areas near the municipal limit of the city. Some of them have taken up jobs in factories, e.g. Digvijay Cement in Ranip, Gujarat Steel Tube in Kali, and small factories situated in the different parts of the suburb. In this area of suburbs, people are more mobile. Besides daily interaction with the city for work, they go to the city for recreation, medical facilities and for special shopping facilities. Due to the change in occupatation and more dependence on the city, a change in behaviour and life-style of the people is noticeable, e.g. possession of radio, scooter, tape recorder, etc. The younger generation gets attracted more to the city for jobs and recreation and they try to dress up in urban style.

(iii) *The Moderately Urban Suburb* has closer contact with the city. This suburb is only 3 km from the city limit and 9 km from the city-centre. This suburb has more metalled roads and consequently more city bus routes (2) and frequency (68) of city buses. A characteristic feature of this type of suburb is the planned residential colonies which have increased both the percentage of the non-agricultural worker (60%), density of population 498 per sq.km) and number of housing societies (10) arc in larger number in 1971 than the rural and moderately rural suburbs. This type of suburb has street lights. The people of this area have to shift to non-agricultural activities because parts of the cultivated land have been acquired by land developers to put into urban uses. In this section of suburbs, the original village sites are no longer entirely inhabited by its traditional residents. There is an infiltration of inmigrants from other parts of the state and the neighbouring states into the village core. The newcomers with relatively low income have settled down in the village by renting rooms from the villagers as they cannot afford high rent in the city.

(iv) *The Urban Suburb* covers areas just across the city limit. The transportation system of this suburb is very much developed than the previous groups. People of this suburb are dependent upon the main city for all activities. In this suburb, most of the cultivated land is acquired by the land developers. So the number of housing society is also very high compared to the other sections of the suburbs. Most of the non-agricultural land is put to urban uses or reserved for future development, so the land value rises abruptly here. Satellite towns have come up and the original village sites are surrounded by their growth. In this suburb, nothing is left of the rural environment except the old village sites. Due to the inmigration of large number of urban population from various parts of Ahmedabad city in suburban housing societies, the density of population (1162 per sq.km) in 1971 and the growth of population during 1961-71 (145%) are very high in this suburb. This type of suburb has more basic amenities in housing societies, like water taps, electricity and metalled roads than the previous one.

TABLE 4.2: The Characteristics of various types of the Suburbs of Ahmedabad

Type of Suburbs / Characteristics of suburbs	Distance in km Core village from the Municipal boundary	Distance in km Mauja limit from the central city	Registered housing societies	AMTS bus routes in 1971	AMTS bus frequency in 1971	Growth rate of population 1961-1971	Density of population per sq. km in 1971	% of non-agricultural workers in 1971	Number of villages
	1	2	3	4	5	6	7	8	9
1. Rural	11.32	16.24	.75			49.05	231.29	36.73	4
2. Moderately rural	6.00	11.24	1.00	.57	9	50.81	308.96	38.64	6
3. Moderately urban	2.83	8.53	10.00	2.15	67.66	53.35	498.17	60.23	7
4. Urban	1.13	6.18	66.25	5.75	144.75	144.75	1162.56	63.43	4

Sources:Ahmedabad Municipal Transport Service, Co-operative Society. District Census Handbook 1961, 1971.

N.B.: Classification is made on the basis of 'Suburbanisation Index'.

This study has concentrated on the classification and types of suburbs to understand the process of transformation of villages from rural to urban form. The framework is based on the changes in landuse, demography and infrastructure. It is used as a base to find out the structural and functional characteristics that are responsible for the formation of the present suburb of the residential area of Ahmedabad.

Fig 4.12

CHARACTERISTICS OF URBAN & MODERATELY URBAN SUBURBS
1981

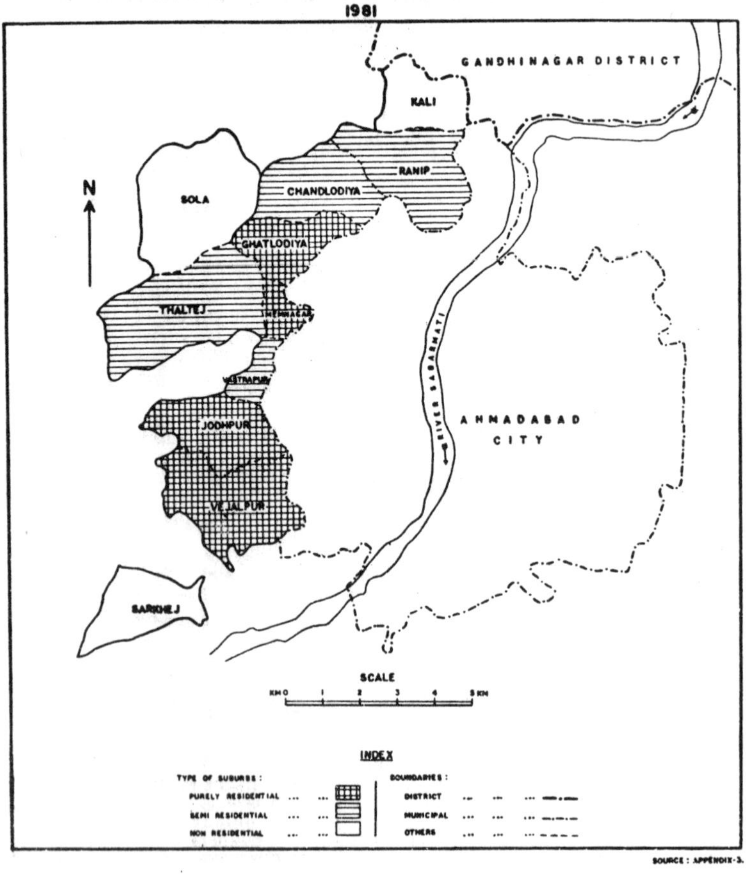

SOURCE : APPENDIX-3.

(b) The Characteristics of moderately urban and urban Suburbs

The moderately urban and urban suburbs of the city are further classified with the help of 'Residential Index' (Figure 4.12). The classification of suburbs by this index is based on the distance from the city centre, percentage of residential area, number of built-up housing societies, bus routes and frequency of the city bus services. The method of classification has been explained in Chapter-III. With the help of 'Residential Index', the moderately urban and urban suburbs are further classified into following three zones, i.e. non-residential, semi-residential and purely residential suburbs (Figure 4.12 and Table 4.3).

The moderately urban and urban suburbs are much desired areas for the establishment of various urban activities because they enjoy the facilities of metalled road, street light, city bus service and also other amenities of the city-centre. So this area gets devoloped both as residential and non-residential suburbs. Characteristics of these suburbs are described briefly in the following section.

71. Ahmedabad City, Its Suburbs and the Study Area

TABLE 4.3: **The Characteristics of urban and moderately urban suburbs**

Characteristics / Type of Suburbs	Distance from the city centre	% of residential areas to total mauja area, 1981	Built-up housing societies 1981	AMTS Bus 1981 No. of routes	AMTS Bus 1981 Frequency	No. of maujas
	1	2	3	4	5	6
1. Non-residential	11.00	22.67	3.32	2	52.33	3
2. Semi-residential	6.64	38.82	30.5	2.5	69.5	4
3. Purely residential	6.02	50.93	53.00	5	137.5	4

Source: Village Panchayat offices and AMTS office.

N.B.: Classification is made on the basis of 'Residential Index'.

(i) *Non-Residential Suburb* of Ahmedabad city is either an industrial (Kali) or commercial node (Sarkhej) and in some cases it is a religious centre (Sola). In industrial or commercial node, the original core villages become living place for bluecollar working people. This influx of low class working pople from various parts of the country and the state has made core villages more crowded. Despite the provision of few urban amenities, e.g. public water tap, street light, etc., the living condition of the core villages is very poor. The deteriorating condition of villages and industrial activities do not allow the growth of new residential area. So the average number of built-up housing societies (3) and percentage of residential area (22.67%) are very low compared to that of semi-residential and purely residential areas.

(ii) *Semi-Residential Suburb* has a mixed type of non-agricultural land use. In this suburb, residential land occupies larger areas (38.82%) and the number of builtup housing societies is also more (30) than the non-residential suburb. In this suburb, the non-agricultural landuse other than residential one is generally occupied by various educational institutions, i.e. Indian Institute of Management, Sardar Patel Institute of Economics and Social Sciences, and Government organisations i.e. railways, food corporation, oil godowns and small factories. As this area is just outside the city limit so residential and non-residential urban activities have grown simultaneously.

(iii) *Purely Residential Suburb* Is just outside the city limit of the residential area of the city. In this suburb, residential landuse is the main landuse. The land value in this area has risen abruptly because the land possesses high potentiality for residential development.

All three suburbs described above are connected with the city-centre and to other parts of the city by city transport service. The majority of the poeple of this area depend on Ahmedabad city for work, education, shopping and recreation.

The purely residential and semi-residential suburbs have been combined to form residential suburbs (Figure 4.12) and this residential suburbs have been selected as the study area.

3. Summary of the Chapter

The river Sabarmati plays an important role in the growth and development of Ahmedabad city and its suburbs. The original city of Ahmedabad was established in the eastern side of the river and till the later part of the 19th century the growth of the city was concentrated in the eastern part of the city. In the earlier part of the 20th century, due to the construction of bridge over the river, larger areas were opened for urban expansion in the west of the river. This expansion in the west of the river was predominantly residential, while industrial and commercial zones are in the eastern side of the river. Due to the formation of new State of Gujarat, large number of people were attracted to the city for rapid industrialisation of the state. As population pressure started mounting within the city limit, new settlements have started developing in the areas which were formally agricultural land. Agricultural lands were sold to residents of the city by the villagers who were the original residents of the existing core villages. New suburban settlements have grown initially near the city limit, then along the transport routes, and then gradually settlements spread all over. But the core villages have remained as "islands" in the midst of these new settlements.

In this study the suburb of Ahmedabad city has been broadly divided into four zones. The rural suburb is the farthest from the city limit which is followed by moderately rural, moderately urban and last is the urban suburb which is just across the city limit. The rural suburb is more or less rural in character with no city bus service and no built-up urban residential houses. But the influence of suburbanisation has been felt by the presence of non-agricultural fallow and daily movement of residents to the city to supply milk. Residents of the rural suburb possess very few urban modern goods. In the moderately rural suburb, few city buses are present and here the agricultural land has been devoted to commercial farming. Residents of this type of suburb have started seeking subsidiary jobs in the city. They have more interaction with the city than with the residents of rural suburb. The moderately urban suburb is very near to the city limit. It has more frequency of city bus service, higher growth of population and higher density than the previous two types of suburbs. In this suburb, original village site is also no longer inhabited by its traditional residents, because inmigrants, with low income, have settled down there. The urban suburb is more or less the extension of physical characteristics of the city itself. This suburb has electricity, water taps, metalled roads and various other basic amenities though for all activities residents of this type of suburb also depend on the city-centre.

The moderately urban suburb and urban suburb are further sub-divided into three divisions: non-residential, semi-residential and purely residential suburb. In the non-residential suburb, most of the non-agricultural land is either industrial or commercial, while in the semi-residential suburb parts of the non-agricultural area have been used for institutional and various government organisations and the rest is residential. In purely residential suburb, all non-agricultural land has been used only for residential purposes.

The maujas of semi-residential and purely residential suburbs have been selected as the study area.

FIVE

THE STUDY AREA

1. Location

The study area is situated along the western city limit of Ahmedabad city. The study area includes eight maujas in the district of Ahmedabad (Figure 5.13).

All maujas of the study area except one, are just across the municipal limit of the city and are growing very fast. The study area covers 37.82 sq.kilometres and have 124, 258 as total population in 1981. The average motorable distance from the core city of Ahmedabad is 7 km. This area is connected with the city-centre by metalled roads and Ahmedabad Municipal Transport Service runs their buses in this area. The landuse of the study area is mainly residential, though in some cases it is mixed with the institutional landuse in the central area (Vastrapur) and industrial landuse in the north (Ranip).

2. Factors of Development

The present study area has been developed in the later part of 1960's. This development is not a continuous one. The southern and the northern most maujas had developed earlier. Here the word "development" has been referred to the rate at which agricultural lands has been changed into non-agricultural use and the rate of construction of residential houses in the form of housing societies in a mauja. The co-operative form of housing societies is the predominant residential structure in the study area as well as in Ahmedabad. The rate of conversion of agricultural land into non-agricultural use in a mauja is influenced by several factors: Location of the mauja, its accessibility, demand for the land, caste and social status of the original residents of the village core and other social and cultural factors.

Fig. 5.13

LOCATION OF THE STUDY AREA WITH RESPECT TO AHMADABAD CITY

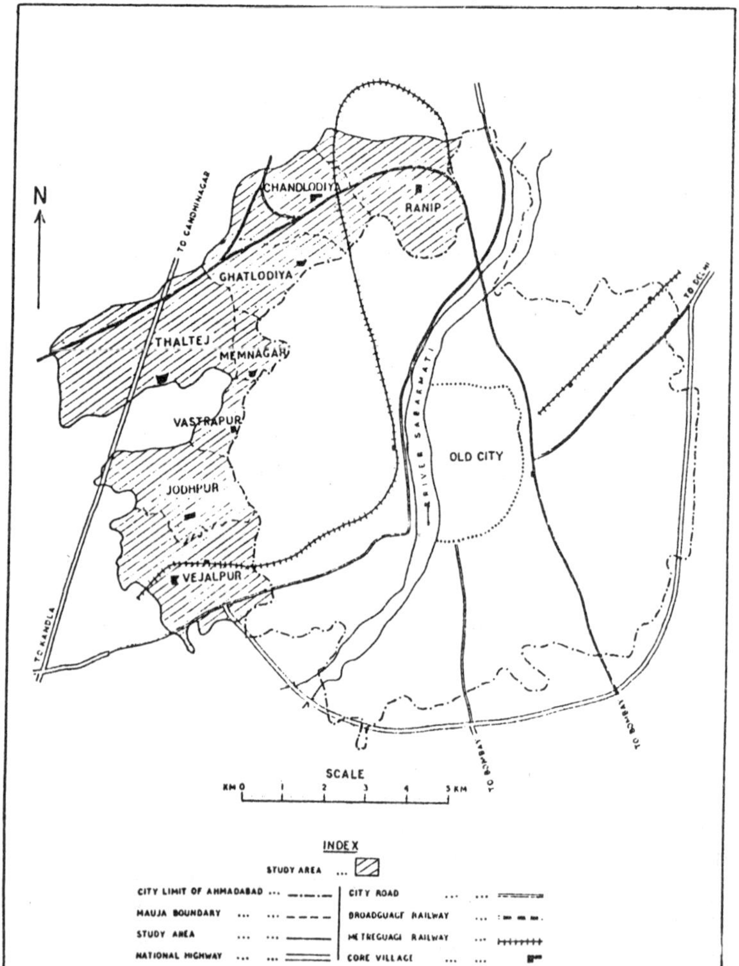

Location of mauja, i.e. its nearness to the city and its accessibility with the city-centre and other parts of the city help to develop a mauja faster. Both these two factors are considered for the development by the urban residents. So, when people start settling down in suburban areas, they prefer the nearest mauja from the city limit and which has high accessibility because both of these factors help them to commute to the different parts of the city easily. These new suburban residents have to keep high interaction with the city for work, education, shopping, etc. Most of them have also friends and relatives in different parts of the city.

Step-wise migration is noticeable in this study. First, residents have settled in areas which were formally the suburbs of the city but now a part of the city, by moving out from the city-core. Some have come from the various parts of the state and country to the core-city or to other parts of the city. Then they have moved to the present suburb where land is cheaper. So, obviously, most of them have friends and relatives in different parts of the city. The growth of population through migration is explained in the next section.

Other factors, i.e. caste of the original residents of the core villages and other socio-cultural factors also help in the suburbanisation of a mauja. The high caste and high socio-cultural status of the people of core village's attract people with high socio-economic status to settle in new suburban areas while maujas with original residents of low caste and low socio-economic status take time to develop.

In the study area, Vejalpur and Jodhpur (Figure 5.13) in the south are highly urbanised maujas. They are followed by Ranip in the north where non-agricultural landuse is of mixed type, i.e. residential and industrial. Vastrapur and Memnagar which have started developing are centrally located and nearest to the city-centre. They started developing before 1960's. Some portions of developed areas of these two maujas have been merged within the city limit. Vastrapur is now one of the least growing maujas in the study area because Vastrapur has the problem of discontinuous growth of residential areas due to the presence of educational institutions within its boundary. Ghatlodiya in the north-central areas has started developing very fast in the later part of 1970's due to the construction of flats by Gujarat Housing Board. Thaltej is the least grown mauja in the study area. The cause of its low growth is its location, i.e. long distance from the city-centre and also social factor like the low caste of the original residents of the core villages. The original residents of Thaltej village are known for social crimes, hence that has made this village less attractive to suburban population. Chandlodiya, anotherleast developed mauja, is developing slowly as it has low connectivity with the city and other maujas.

3. The growth of the study area through migration

It has been noticed that the growth and development of suburbs have been associated with the change of agricultural land into non-agricultural one and also with the growth of population through in-migration of large number of people in this area.

In this study area, most of the residents (83%), except the original people of the core villages (17%), have migrated in this area. These new suburban population have their place of origin not only in Ahmedabad city (11%) but also from different parts of Gujarat State (58%) and other States of India (14%) (Table 5.4). In this section migration of the people of the study area has been analysed in the following manner, (a) by origin of migrant, (b) by stages of settlement, and (c) spatial distribution by origin and stages.

(a) *Origin of residents*

In the study area, few people have migrated from the core-city of Ahmedabad directly. Due to socio-economic reasons, the original residents of the core-city prefer to stay in the core-city itself, mainly in their ancestral house. But gradually due to inmigration, the core-city has become congested and noisy. There is lack of space in ancestral house; so, gradually younger generation with higher level of education, occupation and income has started moving out from the core-city. Some of them have moved and also settled in the suburb directly, but a large number of them has settled first in different areas within the city limit. As the city limit is extending externally (Figure 4.10), a large portion of the present city was previously a suburb. The earlier outmigrants of the core-city have settled in these areas by building spacious houses in peaceful localities, mainly in the western side of the river Sabarmati.

78. Residential Pattern of Suburbs.

TABLE 5.4: **Place of the birth of residents of the study area**

Core City of Ahmedabad	Other parts of Ahmedabad	Study area	Other parts of Gujarat	Other parts of India	Total percentage
11	Nil	17	58	14	100 (Total sample 1500)

Source: Field Survey by the author.

None of the residents (head of the family) has his place of birth in other parts of Ahmedabad (excluding core-city) (Table 5.4). It is mainly due to the fact that other parts of Ahmedabad city have developed only by the expansion of the core-city, so these are not original place of birth for any one. Persons, who have been born there, are either staying in thier own place of birth or may have migrated in the study area as a member of the family.

The influx of a large number of population in the study area has been reflected in the presence of small percentage (17%) of the people in the original core villages. People of the core villages are the original residents of the study area. So, they are the only residents of the study area, who have not shifted from their place of birth. It is mainly due to their traditional culture and lower level of income, education, occupation compared to that of new suburban population. These original residents of core villages live separately in the study areas as an "island" in the midst of new suburban development.

The study area has large number of population (58%) from different parts of Gujarat. It reflects the importance of Ahmedabad city and its suburbs to the people of other parts of Gujarat. Ahmedabad, being the main commercial and administrative centre (till 1970) of Gujarat, a large number of population from different parts of

Gujarat has migrated in this city. A large number of these people engaged in bluecollar job is very poor and they live in the industrial area (eastern part) of the city, and also in the core-city or in the industrial suburb of Ranip. Most of the people from other parts of Gujarat who have settled in the study area are engaged in whitecollar jobs.

210 (14%) heads of families of the sample are from different parts of India other than Gujarat. As this study area is the residential suburbs of Ahmedabad, most of the non-Gujarati people here are engaged in different services in the institutions, situated in the study area or are engaged in managerial posts in commercial organisations in the city.

All residents of the study area, except the original residents of core villages (17%), have shifted from their place of birth to stay in their present locality. The people have changed their residence many times within Ahmedabad city and study area before settling down in the present locality (Table 5.5). 1500 sample residents have been traced for four locations (Table 5.6). Twenty five per cent of the residents of total sample have moved once. while 19% have moved more than twice (Table 5.5). Seventeen per cent of the residents are non-movers who are the original residents of the core villages.

TABLE 5.5: **Change of residence before settling in present locality (in percentage)**

Never	Once	Twice	More than twice	Total percentage
17	25	19	39	100 (total sample 1500)

Source: Field Survey by the author.

There are variations in movement among the residents of different origin (Figure 5.14) (Table 5.6). 67% of the original residents of core-city of Ahmedabad and 60% of the original residents of other parts of Gujarat have moved more than twice within Ahmedabad city and the study area but none of the residents of the present study area, from other parts of India, have moved more than twice (Table 5.6).

Fig 5.14

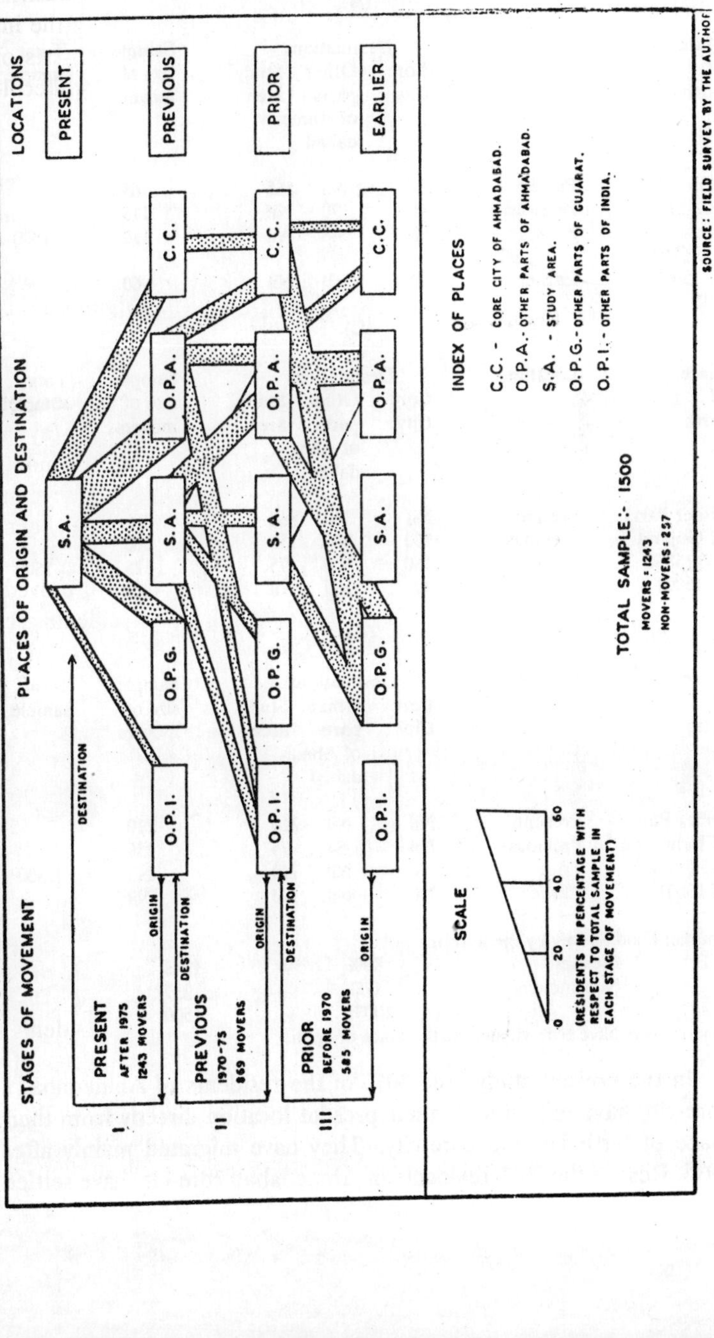
A SCHEMATIC DIAGRAM OF STEPWISE MIGRATION OF THE RESIDENTS OF THE STUDY AREA
STAGES OF MOVEMENT
PLACES OF ORIGIN AND DESTINATION
LOCATIONS
I
PRESENT
AFTER 1975
1243 MOVERS
II
PREVIOUS
1970-75
869 MOVERS
III
PRIOR
BEFORE 1970
585 MOVERS
DESTINATION
ORIGIN
DESTINATION
ORIGIN
DESTINATION
ORIGIN
S.A.
O.P.I.
O.P.G.
S.A.
O.P.A.
C.C.
PRESENT
PREVIOUS
PRIOR
EARLIER
SCALE
0
20
40
60
(RESIDENTS IN PERCENTAGE WITH RESPECT TO TOTAL SAMPLE IN EACH STAGE OF MOVEMENT)
TOTAL SAMPLE:- 1500
MOVERS : 1243
NON-MOVERS : 257
INDEX OF PLACES
C.C. - CORE CITY OF AHMADABAD.
O.P.A. - OTHER PARTS OF AHMADABAD.
S.A. - STUDY AREA.
O.P.G. - OTHER PARTS OF GUJARAT.
O.P.I. - OTHER PARTS OF INDIA.
SOURCE: FIELD SURVEY BY THE AUTHOR

TABLE 5.6: **Sequential location of movers of the study area in relation to their place of birth**

(A)

Place of birth	Locations	Destinations: Core City	Other parts of Ahmedabad	Study area	Sample size of movers	Total sample
Core City of Ahmedabad 165 (11% of 1500)	Present	Nil	Nil	165	165	
	Previous	45	70	Nil	115	
	Prior	62	48	Nil	110	1500
	Earlier	60	Nil	Nil	60	

(B)

Place of birth	Locations	Destinations: Core City	Other parts of Ahmedabad	Study area	Sample size of movers	Total sample
Other Parts of Gujarat 868 (58% of 1500)	Present	Nil	Nil	868	868	
	Previous	150	300	150	600	
	Prior	150	300	75	525	1500
	Earlier	40	185	Nil	225	

(C)

Place of birth	Locations	Destinations: Core City	Other parts of Ahmedabad	Study area	Sample size of movers	Total sample
Other Parts of India 210 (14% of 1500)	Present	Nil	Nil	210	210	
	Previous	Nil	84	75	159	
	Prior	Nil	Nil	Nil	Nil	1500
	Earlier	Nil	Nil	Nil	Nil	

Source: Field Survey by the author

N.B.				
	Movers	83%	=	1243
	Non-movers	17%	=	257
		100%		1500

Non-movers have core villages as the place of birth.

In the present study area 30% of the residents of Ahmedabad's core-city have migrated to their present location directly from their place of birth i.e. the core-city. They have migrated mainly after 1975. Rest of the 70% residents of Ahmedabad core-city have settled in the suburbs in their present location, after changing their residence number of times within the corecity and other parts of Ahmedabad. It is interesting to note that none of the residents from the core-city has changed their house within the study area (Table 5.6). In the previous location, a large (61%) number of residents from the core of Ahmedabad has stayed in different parts of Ahmedabad (excluding core-city). While in their prior location a large (56%) number of these residents have stayed in the core city itself and in their earlier location all of them were only in the core-city. It shows that in the earlier period of migration residents of the core-city moved within the city and nearer to their own relatives. In that period, they did not migrate in the other parts of the city, not only because those areas are not properly developed, but also the residents had very little knowledge and perception about areas outside the core-city. In later periods, when other parts of the city have started developing and the core-city has become a crowded, commercial centre, the residents of the core-city started migrating to other parts of the city and some of them have directly migrated from the core-city to their present location in the suburban study area in the recent years. Many of the direct movers of core-city are Muslim population and they have settled in a particular part (extreme south) of the study area and have formed societies based on religion. They have migrated from the core-city due to the demand for more personal space and peaceful locality but they have concentrated in one place in order to have social security.

In the study area, the largest (58%) number of residents are from other parts of Gujarat. The people from other parts of Gujarat have changed their residence many times within different parts of Ahmedabad city and the study area before settling down in their present location. Only 31% of the residents of other parts of Gujarat have directly migrated in their present location from their pkce of birth. While 9% of them have changed their residence only once within the city and the study area and 34% have changed their residence twice and 26% have changed their residence more than twice (Table 5.5).

The residents from different parts of Gujarat have started migrating in this city since 1960, that is after the formation of the State of Gujarat. At that time the present study area was purely rural around the core-city in rented houses. As the city started growing, these prople also started migrating in newly developed areas, either by building own house or by getting cheap rented house in the prior stage. Those prople from other parts of Gujarat who have migrated in the city relatively in recent years, have migrated in the western part of Ahmedabad city, as the suburb is now more developed. So new inmigrants with relatively high income have settled in other parts of Ahmedabad and the study area directly without settling in the core-city.

24 % of the residents from other parts of India have directly migrated in their present location from their place of birth. All other non-Gujarati residents (i.e.76%) of the study area have changed their location only once within other parts of Ahmedabad city and the study area before settling down in their present location (Table 5.6). It is interesting to note that none of the non-Gujarati residents of the present study area has stayed in the core-city. Most of them, due to their food habit (non-vegetarian), did not get rented house in the core-city and some of them have not preferred the core-city due to its noise, congestion and being far from the work place (educational institutions) which is often in the study area.

(b) *Stages of Settlements*

It is obvious that most of the people of the study area have come from different origins and have gone through different stages of migration within Ahmedabad city and the study area before settling down in their present locality (Figure 5.14). So, in this study, three stages, i.e. present, previous and prior stage of migration of the population of the study area has been studied by tracing individually the movers by origin and destination in matrix form (Table 5.7).

In the present stage of migration, a large number (22%) of people have migrated directly from other parts of Gujarat (Figure 5.14 and Table 5.7A) to the study area.

In this present stage (after 1975), the study area has developed a lot and relatively cheap rented houses are available. Moreover, the perception of the core-city and of suburbs is the same for residents who have come from long distance. But residents of the core-city or who have come from long distance. But residents of the core-city or other areas of the city did not like to move to suburbs directly because it is an unknown underdeveloped area which has not created any image in their minds. So, people have started migrating direclty in the study area from other parts of Gujarat and India (Table 5.7A). At this present stage, a large number of people have also shifted their residence within the study area. These people in their earlier stages of migration have shifted in suburbs mainly in rented houses. Gradually, they have changed their residence within the study area, either by building their own house or by getting better house on rent.

In their previous stage (Stage II) of migration, i.e. during 1970-1975, a number of the residents has migrated in the study area from other parts of the study area, Ahmedabad, Gujarat and India (Table 5.7B). At this stage the study area was developing rapidly into a true residential suburb which has attracted large number of migrants to settle in a peaceful locality where land-value is also cheap. In this stage very few (6%) people have moved within the core city. But some (43%) of them have shifted within other parts of the city or migrated to the other parts of the city from core-city, other parts of Gujarat and India (Figure 5.14).

TABLE 5.7: **Origin-destination of movers by stages of migration (In percentage)**

(A) Present Stage (I) : Population who have shifted once (after 1975)

Previous origin	Present	Destination: Core city	Destination: Other parts of Ahmedabad	Destination: Study area
1.	Core city	Nil	Nil	19
2.	Other parts of Ahmedabad	Nil	Nil	36
3.	Study area	Nil	Nil	19
4.	O.P. Gujarat	Nil	Nil	22
5.	O.P. India	Nil	Nil	4

Total sample ofmovers = 1243

i.e. 83% of total sample selected (1500) — 100% (1243)

(B) Previous stage (II) : Population who have shifted twice (1970-75)

		Previous	Destination: Core city	Other parts of Ahmedabad	Study area
Prior					
	1.	Core city	6	15	2
	2.	Other parts of Ahmedabad	Nil	16	19
ori-gin					
	3.	Study area	Nil	Nil	14
	4.	O.P. Gujarat	Nil	2	7
	5.	O.P. India	Nil	10	9

Total sample of movers = 869
i.e. 58% of total sample selected (1500) 100% (869)

(C) Prior stage (III) : Population who have shifted more than twice (before 1970)

		Prior	Destination: Core City	Other parts of Ahmedabad	Study area
Earlier					
	1.	Core city	7	10	Nil
	2.	Other parts of Ahmedabad	Nil	30	Nil
ori gin					
	3.	Study area	Nil	Nil	Nil
	4.	O.P. Gujarat	19	20	13
	5.	O.P. India	Nil	10	9

Total sample of movers = 585
i.e. 39% of total sample slected (1500) 100% (585)

Source: Field Survey by the author

In the prior stage (III) of migration, i.e. before 1970, a large number of people have migrated to the study area from other parts of Gujarat. Many of them were construction workers, who were engaged in different construction works in the study area. In the prior stage, most (60%) of the movers have moved either within other parts of the city or migrated there from other places (Table 5.7c). In the prior stage, many people have migrated to the core-city also from other parts of Gujarat (Figure 5.14 and Table 5.7C).

In all stages of migration, most of the residents have migrated from other parts of Ahmedabad City. None of them have migrated from the other parts of the city to the core-city. It shows that residents, who have once migrated into the other parts of Ahmedabad city, have either moved within their locality or preferred to move to suburbs than to migrate to their previous locations of the core-city.

A generalised spatial model of state-wise migration of the residents of the study area has been formulated from the above mentioned analysis by origin of movers and by number of moves that they have made. This model has given an idea of origin of movers in step-wise process of migration in the suburb by tracing 1500 people individully (figure 5.15). In this model, the core-city and the whole study area have been considered as a single unit in all locations. The other parts of Ahmedabad city (excluding core-city) has been divided into two units, i.e. western part with new settlements and eastern part with industrial areas of the river Sabarmati. While other parts of Gujarat has been sub-divided into four groups, i.e. north (Banaskanta, Mehsana, Sabarkantha): east (Ahmedabad, Panchmahala districts):south (Kheda, Vadodara, Bharuch, Surat, Valsad and Dangs districts): west (whole of Saurashtra and Kutch) and other parts of India has also been divided into four groups, i.e. north (Jammu and Kashmir, H.P., Punjab, Haryana, Delhi and U.P.): east (W.B., Bihar, Orissa, Assam and North Eastern States): south (A.P., Tamil Nadu, Kerala, Karnataka) and west (Maharashtra, Rajasthan and M.P.).

In present stage, the original location of most of the residents of the study area is on the western part of Ahmedabad city (Figure 515A). In this stage, most of the residents from other parts of Gujarat have migrated from the eastern part, while in the case of other parts of India it is from South. In other two earlier stages (previous and prior), most of the present residents had their place of origin in the western part of Ahmedabad city, western Gujarat and Western India (Figure 5.15 B,C)

The residents living in the western part of Ahemdabad city have a clear perception about the study area, because the study area is the nearest to the western part of the city than the other parts. So, a number of residents of the present study area were living in the western part of the city previously. This is also true about the people from the western part of India.

Fig. 5.15

A GENERALISED SPATIAL MODEL OF STEPWISE MIGRATION: ORIGIN OF SUBURBAN POPULATION

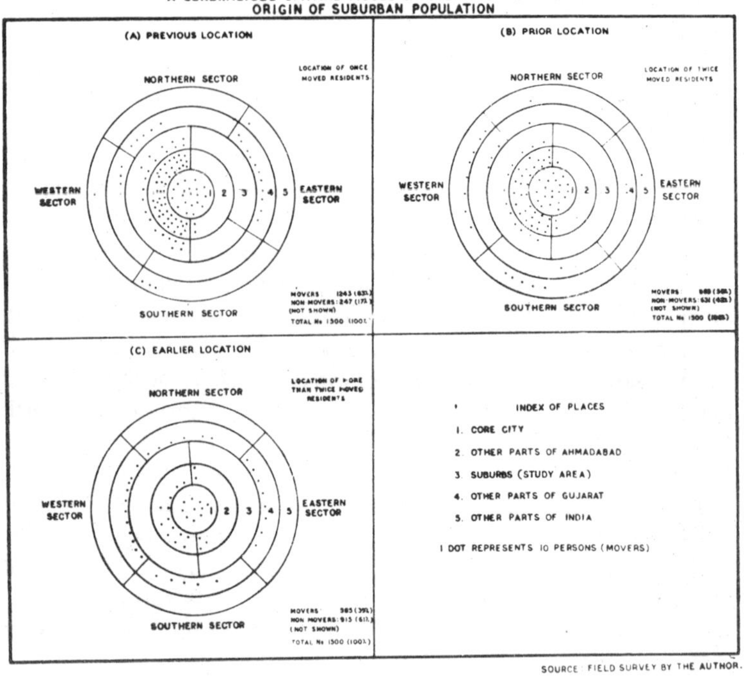

This generalised model of step-wise migration of suburban population can be applied to most of the Indian suburbs, mainly in suburbs of pre-industrial cities. This model helps us to understand the composition of population in the study area and also the growth of the study area in different stages with respect to the origin of residents with sequential movements.

4. Housing Structure

The study area is no just merely a complex organisation of social life. The agglomeration of many types of houses in the study area makes the physical landscape also a complex system. The study area, like many other suburbs of pre-colonial cities of India, shows

Fig. 5.16

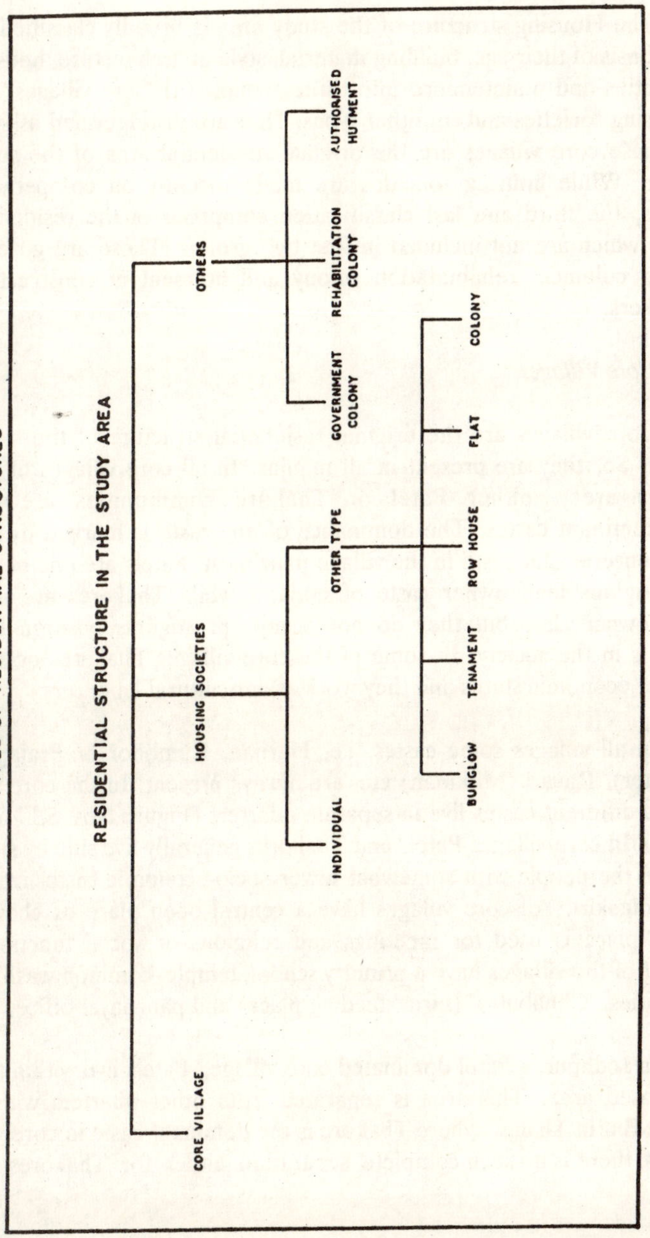
TYPES OF RESIDENTIAL STRUCTURES
RESIDENTIAL STRUCTURE IN THE STUDY AREA
CORE VILLAGE
HOUSING SOCIETIES
OTHERS
INDIVIDUAL
OTHER TYPE
BUNGLOW
TENAMENT
ROW HOUSE
FLAT
COLONY
GOVERNMENT COLONY
REHABILITATION COLONY
AUTHORISED HUTMENT

difference in the types of housing structure (Figure 5.16) which can be grouped into meaningful categories of similar character.

The Housing structure of the study area is broadly classified on the basis of their age, building material, style of architecture, housing facilities and maintenance into three groups: (a) Core villages, (b) housing societies and (c) other types. They are distinguished as such because core villages are the original residential area of the study area. While housing societies are made recently on co-operative basis, the third and last classification comprises of the residential type which are not included in first two groups. Those are government colonies, rehabilitation colony and hutment of construction workers.

(a) *Core Villages*

Core villages are the original residential structure of the study area. So, they are present in all maujas. In all core villages of the study area, either Patel or Thakore communities are the predominant castes. The dominance of any caste is marked by the presence of that caste in the village panchayat. Patels are one of the prestigious land owner caste of Gujarat while Thakores are also landowner class but they do not occupy prestigious position like Patels in the society. In some of the core villages Thakores occupy lower economic status and they work as agricultural labourers.

In all villages some castes, i.e. Harijan, Kumbhor or Prajapati (potter), Ravari (Milkman) etc. are always present. In the core villages different castes live in separate quarters (Figure Nos. 5.17 and 5.18). In core villages Patels and Thakores generally live side by side, while the people with somewhat lower socio-economic caste live in the outskirt. All core villages have a central open place or chawk. This place is used for meetings and religious or social functions. Most of the villages have a primary school, temple, common washing facilities, "Chhabutra" (birds' feeding place) and panchayat office.

In Jodhpur, a Patel dominated core village, Patels live within the enclosed area. This area is separated from other quarters with a gate. But in Thaltej, where Thakore is the dominant caste in core villages, there is no such complete separation either for Thakores or

Fig. 5.17

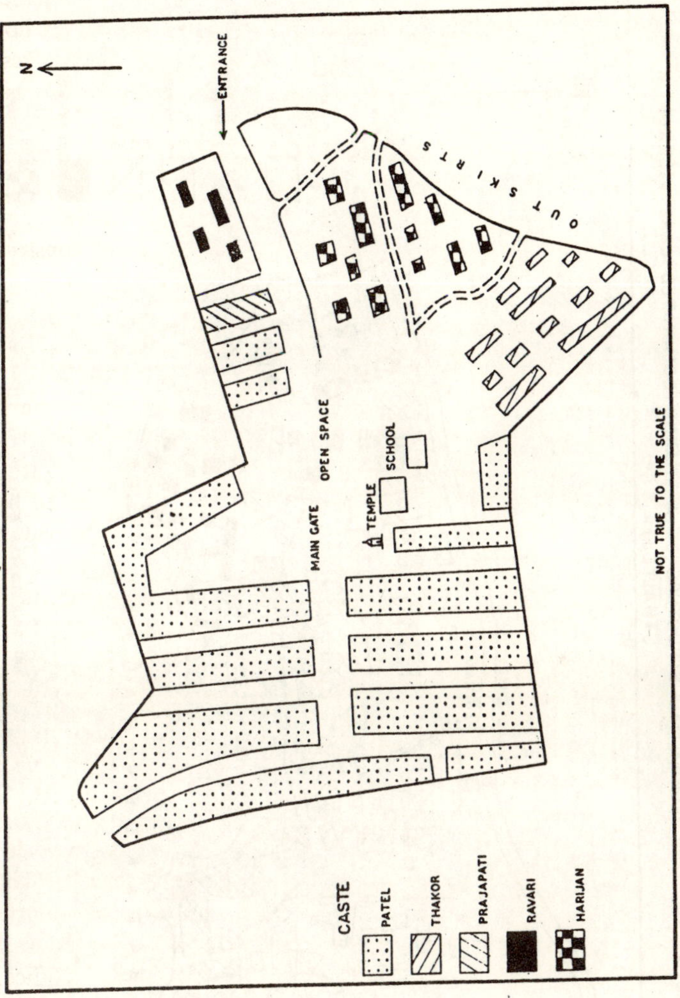
A SKETCH OF THE RESIDENTIAL STRUCTURE OF A PATEL DOMINATED CORE VILLAGE
(JODHPUR)
N
ENTRANCE
OUT SKIRTS
OPEN SPACE
MAIN GATE
SCHOOL
TEMPLE
NOT TRUE TO THE SCALE
CASTE
PATEL
THAKOR
PRAJAPATI
RAVARI
HARIJAN

Fig. 5.18

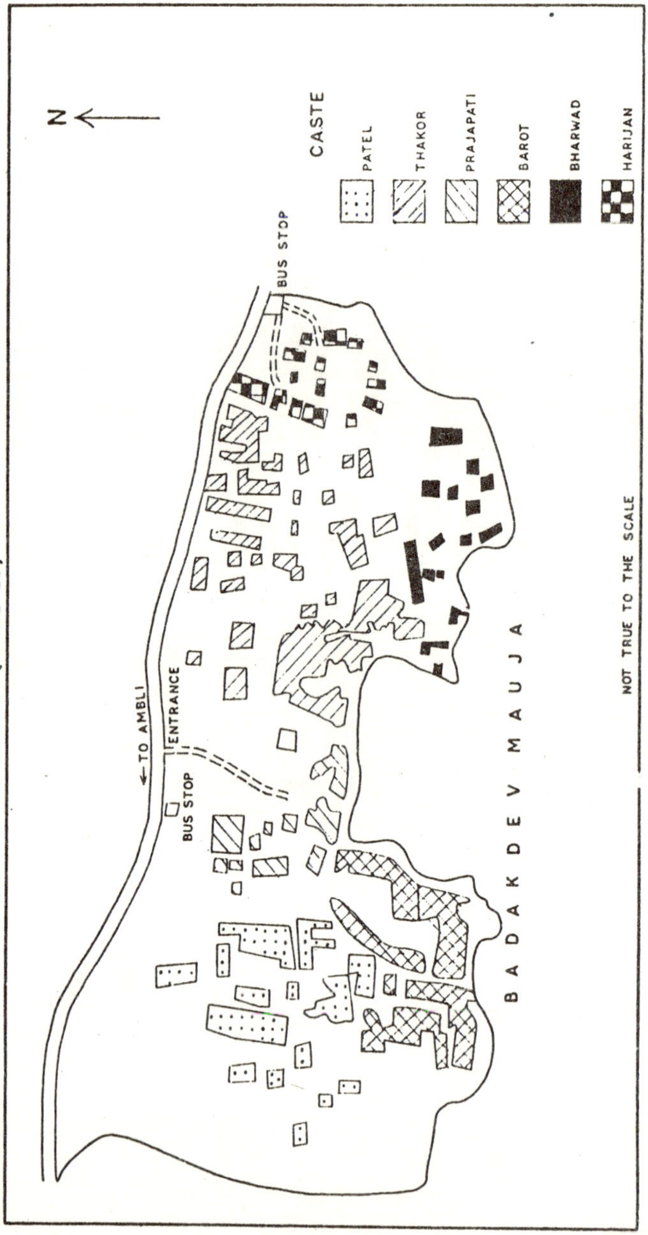
A SKETCH OF THE RESIDENTIAL STRUCTURE OF A THAKOR DOMINATED CORE VILLAGE
(THALTEJ)
N
CASTE
PATEL
THAKOR
PRAJAPATI
BAROT
BHARWAD
HARIJAN
BUS STOP
TO AMBLI
ENTRANCE
BUS STOP
BADAKDEV MAUJA
NOT TRUE TO THE SCALE

for Patels. Vejalpur, Vastrapur and Thaltej are dominated by Thakore caste while Jodhpur, Memnagar, Ghatlodiya and Chandlodiya core villages are dominated by Patels. Houses of the core villages are built up of bricks, tiles, stone, leaves etc. They are devoid of any facilities like separate toilet, bathroom, kitchen, water tap, etc.

Most of the residents of these core villages have sold their agricultural land to housing societies or institutions but they have bought agricultural land in the rural suburb. So most of them are still practising traditional agricultural occupation. But some of them have changed their occupation to non-agricultural activities like business, office jobs, etc. after the sale of land.

These villages were surrounded with agricultural land previously. But, now the agricultural land is sold, so the core villages are surrounded with new suburban housing societies. As a consequence, these rural communities have remained as "islands" in new suburban environment.

(b) *Co-operative housing societies*

Co-operatives housing societies are the prominent housing structure in the study area. This type of housing structure is present in all maujas. In this type of housing structure a large number of houses are present in a co-operative society. The number of houses in a society varies from ten to more than 200. Vejalpur, an earliest developed mauja, has the largest (115) number of housing societies in the study area and it has been followed by Ranip (72), while recently developed maujas like Vastrapur and least developed maujas like Thaltej have few housing societies (Figure 5.19). In this study, the growth and development of a mauja has been analysed in relation to the presence of housing societies. So, obviously, developed maujas will have more housing societies than the underdeveloped maujas in the study area. There are two types of co-operative housing societies in the study area. In one type, the co-operative societies are established to acquire sites for members who want to own house. In this type, co-operative society only sells the land to the members. This type of co-operative housing socieity is only 4% of the total housing societies of the study area. They are existing only in Vejalpur and Jodhpur. This type of socieities are termed as 'Individual Bungalow' in the study area.

Fig. 5.19

DISTRIBUTION OF VARIOUS HOUSING STRUCTURES (HOUSING SOCIETIES) 1981

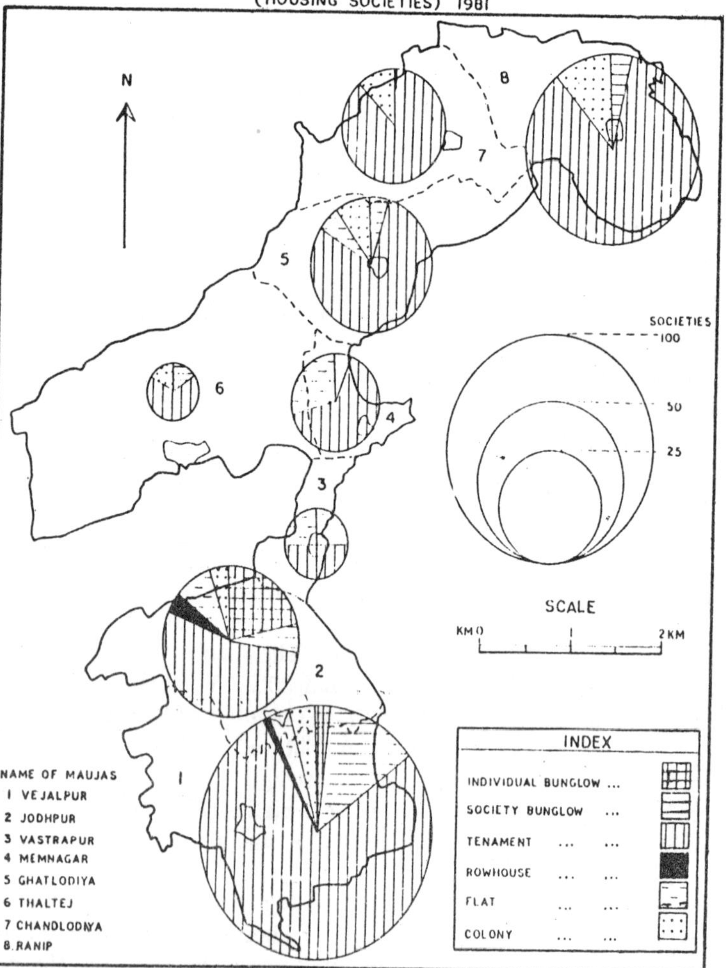

In the second type of societies, the co-operative organisation assigns dwellings to their members. Co-operatives of this sort provide facilities like electricity, water and sewerage. This type of society are distributed all over the study area and have five types of housing structure (Figure 5.16). In this type of society, all houses in a single society have similar structure, built up area, number of rooms, etc. when the possession has been given.

The structure of the houses of co-operative housing societies varies over time and space. In the initial stage of development when enough land was available in the suburb, low spacious bungalows were visible. Then tenements were common. Gradually houses are being built, using the vertical space, e.g. row houses and flats are now new trend in the housing structure.

Bungalows, built up by these housing societies, are generally detached, one storied house with a provision for the vertical expansion. Though these bungalows are present in all maujas, except Chandlodiya, they occupy only 8% of total number of housing societies.

Tenements are the most common type of residential units in the study area. They are semi-detached houses in which two dwellings are joined by a common wall. These are then twin bungalows in which residents share a common wall. These houses are also single-storied with a provision for vertical expansion.

Row houses are recent phenomenon in the housing types of suburbs. These are two-storied houses where every house has a common wall on both sides. Every house has separate open space. This type of house is seen mainly in Vejalpur and Jodhpur. It occupies only 1% of the total houses of the study area.

Flats are on the whole one or two bed room units with living-cum-dining room. On the whole, they are three-storied buildings, though there is also a pattern of having parking space on the ground level with three levels above.

Colonies are somewhat modern type of 'chawls' which are built to house industrial workers in the industrial sector of the city. These are arranged in lines of single-storied houses with one room and kitchen. Main difference of this type of house from the 'chawl' is that it has separate toilet and water tap in each house.

There is a great variation within the same type of residential structure of societies regarding total built-up area, number of rooms etc. So, it is not possible to compare between different types of residential structures of the housing societies. In the following chapters housing societies will be considered as a single type to be compared with that of core villages in a mauja. Differences between the housing characteristics of various maujas will also be analysed.

(c) *Other type of houses*

These are neither core villages nor housing societies. These are Government quarters, colonies or hutments.

Government colony has been built by Gujarat Government in Vastrapur for government and semi-government employees. These are four-storied flats which are grouped on the basis of income status of the occupants. These buildings have all provision for structural amenities and the demolition and additional construction of any part or whole of the building is prohibited.

There are two *rehabilitation colonies* in the study area and both of them are situated in Vejalpur. One of these colonies was built for backward Thakorda community, while another colony, called Sankalitnaga was built to accommodate families belonging to different caste and communities who were uprooted from their own slums on the bank of river Sabarmati during the heavy floods of 1973. Both the colonies have been developed by private, semi-government and government agencies. Houses in these colonies are of one room without any separate toilet or water tap.

Hutments are not very common in the study area. There is one authorised hutment in Jodhpur. Other than this, there are some patches of unauthorised hutments. The slum in Jodhpur was initially an illegal occupation on unused land. This area is gradually occupied by the people who came from villages to work as daily labourer in construction sites. These slum dwellers are the poorest in the study area and their dwellings are poor, devoid of any sanitary facilities.

5. Summary of the Chapter

The maujas of semi-residential and purely residential suburbs have been selected as the study area. The study area has eight maujas of which Vejalpur, Jodhpur, Ranip and Ghatlodiya are very much urbanised with the growth of new houses while Memnagar, Vastrapur, followed by Chandlodiya and Thaltej are less urbanised. The level of urbanisation has been measured by the rate of conversion of agricultural land into non-agricultural use which depends on various factors, e.g. nearness to the city, accessibility, need for personal space by residents, socio-economic status of the residents of the core village. Most of the residents of the study area have migrated in this area. These new suburban population have their place of origin not only in Ahmedabad city, but also from different parts of Gujarat State and other parts of India. All residents of the study area, except the original residents of core villages, have shifted from their place of birth to stay in their present locality. The people have changed their residence many times within Ahmedabad city and the study area before settling down in the present locality. In this study, three stages i.e. present, previous and prior stage of migration of the population of the study area has been studied by tracing individually 1500 sample at four locations. In all stages of migration, most of the residents have migrated from other parts of Ahmedabad city. None of them have migrated from the other parts of the city to the core-city. It shows that residents, who have once migrated into the other parts of Ahmedabad city, have either moved within their locality or preferred to move to suburbs, than to migrate in their previous locations of the core city. A generalised spatial model of step-wise migration of the residents of the study area has been formulated. This model has given an idea of origin of step-wise migration of suburban population on space. The study area has mainly three types of residential structures, i.e. core villages, which are the original residential structures existing prior to the development of the suburb, housing societies which have been built due to suburbanisation and the rest is other type, including the government colony, hutments of construction workers.

SIX

MORPHOLOGICAL GROWTH OF THE SUBURBS OF AHMEDABAD

The residential pattern of suburbs is very much associated with changes in morphological aspects - land use, its demographic composition and its infrastructure. An analysis of different aspects of morphology in suburbs helps us to understand the processes that have been responsible for the growth of the study area.

The pattern of morphological aspects of suburbs is characterised by complex processes. As the pressure of demand on urban land increases in the form of higher land value and more intensive use of urban land, city limit and the distribution of population get extended to suburbs. This increase in demand leads to the increase of both non-agricultural and urban landuses in suburbs. Along with this process of change in landuse, landvalue and density of population tend to increase. Individual mobility increases with the expansion of the city and the improvement of connectivity. The change in land value is, therefore, the first indicator which is related to the growth of the suburbs. The following section of this Chapter deals with the processes relating to the distribution and change of landuse and the factors that control the distribution and growth of land value.

1. Land Use

(a) *Process of change*

The agricultural land which are located adjacent to the city boundary along the arterial roads or near the already developed nodes in suburbs, are of great demand for residential development and this gets converted first. In the first stage, when the land has low demand for urban use, speculators start buying land and initial suppliers are farmers who hold most land in suburbs. Farmers normally prefer to sell land to a person who can buy the whole plot of land. That is not possible for the individual person in most of the cases. So, initially land is sold to speculators who then sell it to individual persons. Speculators control the change in landuse by holding or releasing land at various points of time. In the initial stage, no infrastructural various points of time. In the initial stage, no infrastructural facilities are available. In that stage speculators release only small land at low value and to some people low price is very much attractive. So they start settling in an undeveloped area of suburbs, generally located near the city limit.

In the study area, the conversion of agricultural land to non-agricultural use has been stopped after the introduction of the Urban Land Ceiling and Regulation Act of 1976. Before the introduction of this Act, section 65 of Bombay Land Revenue Code, 1879 lays rules and regulations regarding the procedure for the conversion of agricultural land. In areas outside the municipal limit, District Development Officer (DDO) used to grant permission regarding the conversion of land. The government has introduced many acts to control widespread conversion of agricultural land for speculation. Gujarat Government also introduced a ban on the conversion of land within five km from the periphery of the city limit. Government also has introduced Urban Vacant Land Act (1972-74) to control land speculation. According to this Act, non-agricultural land cannot be kept vacant for more than five years after its conversion from agricultural land. But speculators often pay fine and keep the land vacant for more than fifteen years in some cases, till they get profitable price for it. As a result, land remains without any use for several years.

(b) *Distribution of non-agricultural landuses*

The distribution of landuses in the decades of 1961-1981 brings out clearly the change in the landuse pattern of the study area. The uneven distribution and change in landuse, rate of transfer of agricultural land to non-agricultural land, of suburbs of Ahmedabad are visible on the maps of distribution of landuse 1961, 1971 and 1981. Agricultural land is divided into two broad groups, i.e. cultivated land and agricultural fallow. An agricultural land is kept fallow for many purposes, e.g. to get natural fertiliser or to transfer it into non-agricultural land. A certificates of 'no objection' clearance from the government is needed to transfer an agricultural land into non-agricultural one. In case of non-agricultural landuse, non-agricultural fallow has a potential to change into residential, industrial or commercial landuse. In maujas, where non-agricultural fallow covers large areas, those maujas have scope for further urban development. Some maujas, with more occupied residential area or industrial or commercial landuse than non-agricultural fallow, have reached their peak of development and have little scope for further urban expansion.

In 1961, the study area was totally an agricultural area with sites of villages. Most of its area was either cultivated land or agricultural fallow in each mauja. These non-agricultural fallow, which were kept as speculative land, were situated near the city limit or near the main transport routes. But non-agricultural land in the form of industries, railways and government institutions were present in Ranip in the extreme north, some brick factories and educational institutes in Thaltej in the extreme west and also educational institutes in Vastrapur in the centre of the study area; otherwise the overall picture of the study area in 1961 is more or less uniform, with core villages as sole residential areas in the midst of vast stretches of agricultural land.

In 1971, the picture has changed with the emergence of more non-agricultural fallow and new residential areas in Vejalpur and Ranip. An increase in non-agricultural fallows all over the study area, indicates the increasing demand in the urban landuses in those areas. The variation in the increase of non-agricultural land also indicates that the people of core villages did not sell their land at the same time. Maujas where people sold their land for non-agricultural purposes initially like Ranip in the extreme north and Vejalpur in the extreme south had many new housing societies in 1971. Not only the non-agricultural fallow had increased but agricultural fallow had also increased in 1971, which indicates that farmers would sell these land for urban use in near future.

In 1981, the picture has further changed, with the decrease of the cultivated land and an increase of residential and non-agricultural fallow in all maujas (Table 6.8). But it is interesting to note that, till 1981, residential areas are located in patches nearer to the city limit and on main roads. Ranip and Vejalpur, where residents of core villages have started to sell their land to the people of the city in the late 60s, have reached their peak in this year. Due to several reasons like poor accessibility, distance from the bus routes, city limit, industrial landuse etc., there is less scope for further residential expansion in these two maujas. So the non-agricultural fallow occupies very small percentage of total non-agricultural area in these two maujas in 1981 (Table 6.8).

In Jodhpur mauja, due to the presence of Patel caste in the core village, a large number of suburban population with high socio-economic status has been attracted for settlement. The percentages of occupied residential area and non-agricultural fallow are very high. Jodhpur has developed a lot in the last two decades and there is still scope for urban expansion. In Memnagar and Vastrapur, in the centre of the study area, the scope of urban expansion is also high. Although Vastrapur is the nearest to the city-centre, the percentage of residential land is low here, due to the presence of vast institutional land within its limit. In Memnagar, the percentage of the non-residential area is higher than that of the neighbouring Vastrapur but lower than Ghatlodiya in the north. Low percentage of non-agricultural land in Memnagar is due to the poor accessibility and poor availability of infrastructural facilities in this mauja. Both of these maujas, i.e. Vastrapur and Memnagar, have large amount of non-agricultural fallow which have been left vacant for more than a decade due to the low demand of urban land in this area. But gradually the demand of urban land is also increasing. Ghatlodiya, situated in the north of Memnagar, has a high percentage of residential area (22%) and there is also scope for future urban expansion which is possible due to the presence of non-agricultural fallow (11%). In Ghatlodiya, the urban growth is mainly due to the presence of Gujarat Housing Board flats near its city limit. Due to the location of these flats, accessibility and infrastructural facilities have been developed in this area in the last decade. This develop

TABLE 6.8: **Distribution of landuses (Percentage of total area)**

A - 1961

Name of Maujas	Type of Landuse	1 Cultivated land	2 Agricultural fallow	3 Occupied residential area	4 Non-agricultural fallow	5 Other land uses	6 Total percentage	7 Total Area in sq. Km.
1.	Vejalpur	78	13	3	3	3	100	7.24
2.	Jodhpur	78	13	2	3	4	100	5.11
3.	Vastrapur	78	14	3	2	3	100	1.13
4.	Memnagar	79	14	3	2	2	100	1.89
5.	Ghatlodiya	78	12	3	2	5	100	4.36
6.	Thaltej	80	10	4	3	3	100	8.41
7.	Chandlodiya	85	5	3	2	5	100	4.00
8.	Ranip	63	10	4	3	20	100	5.68
	X	77.13	11.38	3.1	2.5	5.88		

Contd.

(Table 6.8 Contd.)

B - 1981

Name of Maujas	Type of Landuse	1 Cultivated land	2 Agricultural fallow	3 Occupied residential area	4 Non-agricultural fallow	5 Other land uses	6 Total percentage	7 Total Area in sq. Km.
1.	Vejalpur	40	10	23	3	18	100	7.24
2.	Jodhpur	40	11	27	12	10	100	5.11
3.	Vastrapur	42	10	15	10	23	100	1.13
4.	Memnagar	41	15	20	19	5	100	1.89
5.	Ghatlodiya	48	9	22	11	10	100	4.36
6.	Thaltej	64	9	10	5	12	100	8.41
7.	Chandlodiya	60	13	13	6	8	100	4.00
8.	Ranip	28	11	21	9	31	100	5.68
	X	46.13	11.00	18.75	9.75	14.98		

Source: Village panchayat offices

ment created a housing boom in this area in the last decade and most of the societies are situated near the city limit.

Thaltej, in the extreme west, is the least urbanised mauja of the study area. There are various physical and social factors like long distance from the city limit, poor accessibility, poor facility of infrastructure and low social status of the population of the core village, which have made this mauja least attractive. All housing societies of this mauja are situated far from the core village but are located near the main road. The low percentage of non-agricultural fallow also indicates that there is little scope for the development of residential area in the near future. Chandlodiya, situated in the north-east of Ghatlodiya and west of Ranip, has higher percentage of residential area than that of Thaltej but lower than all other maujas. The low rate or urbanisation in Chandlodiya is mainly due to the poor accessibility. Till 1979, there was no metalled road connecting this mauja with the city. The mauja as a whole has only one metalled road. So most of the housing societies are situated near it. So, due to the poor accessibility and poor infrastructural facilities, there is little hope of future development which has been indicated by the low percentage of non-agricultural fallow in this mauja. Thus in the study area, distribution of non-agricultural land uses has been controlled by physical and social factors.

(c) *Growth of residential land in 1961-1981*

Variation in the distribution of landuses in the last two decades (1961-1981) has shown that residential land has increased a lot (504.3%) in the study area. This growth is not same in all maujas of the study area (Figure 6.20).

This residential area includes not only the built-up area but also the open space adjoining it. Jodhpur mauja, in the south, has a very high growth rate of residential area. In Jodhpur, number of societies of individual bungalow is the highest in the study area (Figure 5.19). These bangalows, with bigger built-up areas and large open spaces around them, have increased the residential area of this mauja. Though Vejalpur has the largest number of housing societies, it does not have residential area as high as Jodhpur, because houses in Vejalpur are not as big as the houses in Jodhpur and are devoid of open space.

In Ghatlodiya, in the north-west, the housing boom has taken place in the later part of 1970s due to the availability of good infrastructural facilities, recent development of accessibility, higher growth of residential area during 1961-1981.

Fig 6.20

GROWTH OF RESIDENTIAL LAND 1961-1981

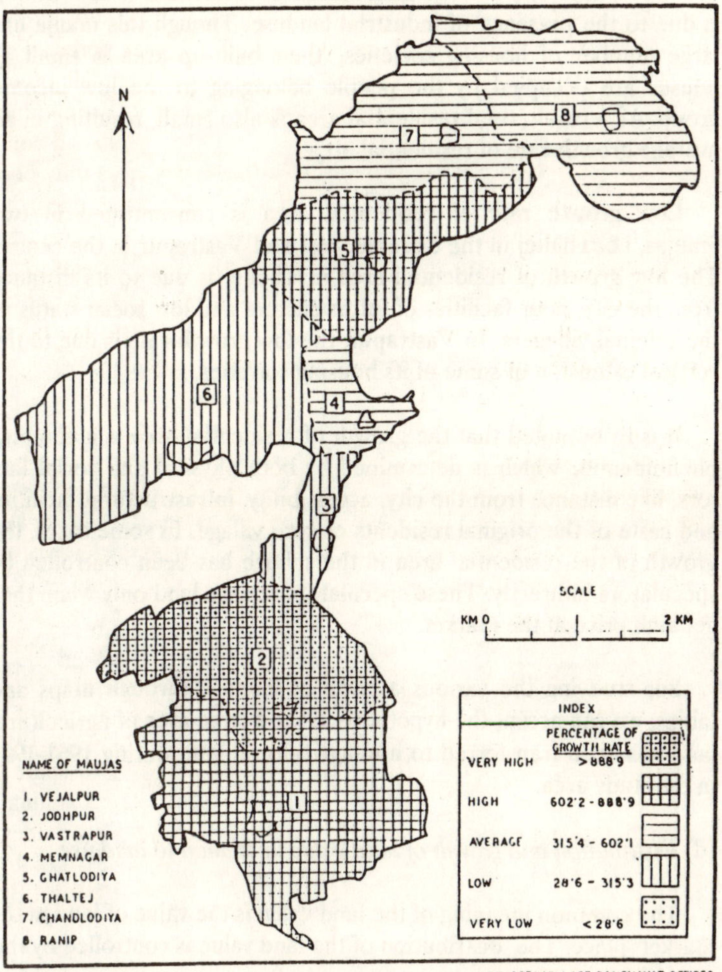

SOURCE: VILLAGE PANCHAYAT OFFICES.

Memnagar in the central part, Chandlodiya and Ranip in the extreme north-west and north respectively, have the growth of residential area in the average group. In Memnagar and Chandlodiya, the growth rate has not increased due to the poor accessibility of these two maujas. In Memnagar, on the other hand, low growth rate of residential area has been recorded, due to the presence of large number of flats, because here residential area is referred to only the horizontal extension, and not the vertical extension. So, Memnagar has higher vertical growth rate of residential landuse which has not been recorded. The average growth rate of residential area in Ranip is due to the presence of industrial landuse. Though this mauja has large number of housing societies, their built-up area is small as houses are occupied by the people belonging to the low income group. As a result, total residential area is also small, resulting in an average growth-rate of residential area.

Low growth rate of residential area is concentrated in two maujas, i.e. Thaltej in the extreme west and Vastrapur in the centre. The low growth of residential area in Thaltej is due to its distance from the city, poor facilities of infrastructure and low social status of the original villagers. In Vastrapur, the low growth rate is due to the vertical extension of some of its housing societies.

It is to be noted that the growth of residential area is a complex phenomenon, which is determined by both physical and social factors, like distance from the city, accessibility, infrastructural facilities and caste of the original residents of core village. In some cases, the growth of the residential area in the suburb has been controlled by speculators indirectly. These speculators release land only when they get high price at the market.

So, studying the various aspects of landuse through maps and tables, we can accept the hypothesis 2, i.e. a large area of agricultural land has been transferred to non-agricultural land during 1961-1981 in the study area.

(d) *Distribution and growth of land value in relation to land use*

The common meaning of the land value is the value of land in the market place. The distribution of the land value is controlled by the

Fig. 6.21

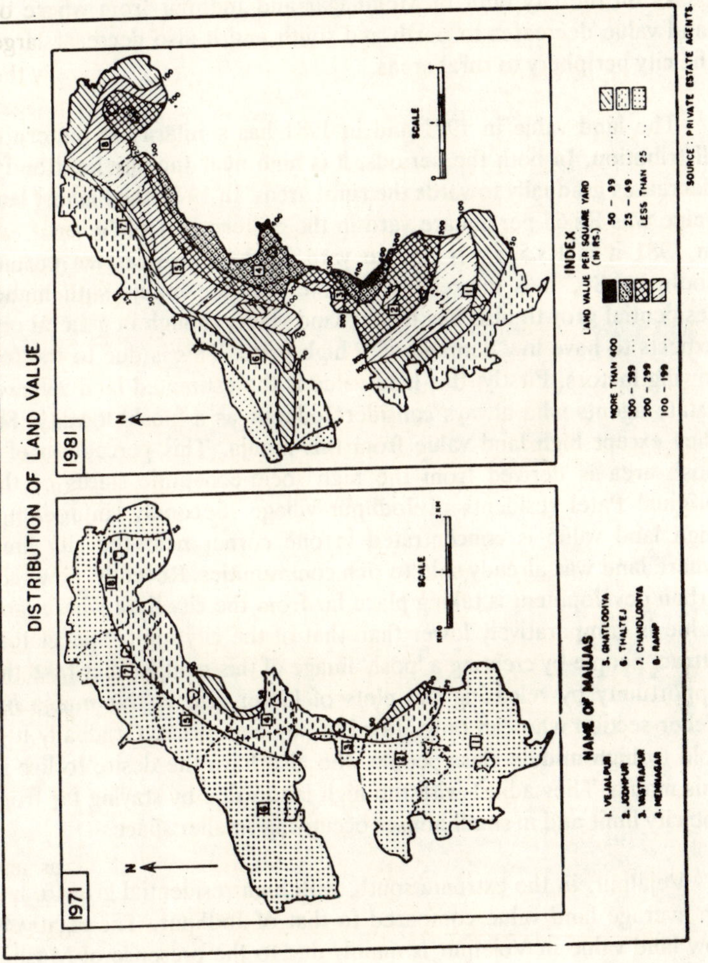
DISTRIBUTION OF LAND VALUE
1971
1981
N
SCALE
NAME OF MAUJAS
1. VEJALPUR
2. JODHPUR
3. VASTRAPUR
4. MEMNAGAR
5. GHATLODIYA
6. THALTEJ
7. CHANDLODIYA
8. RANIP
INDEX
LAND VALUE PER SQUARE YARD (IN RS.)
MORE THAN 400
300 - 399
200 - 299
100 - 199
50 - 99
25 - 49
LESS THAN 25
SOURCE : PRIVATE ESTATE AGENTS.

type of landuse and by its location. This is reflected in the pattern of distribution of land value of the study area (Figure 6.21). An isometric map of land value presents a detailed picture of the land vlaue of the study area (Figure 6.22). In this map, there are two peaks on the city limit of Memnagar and Jodhpur from where the land value decreases to north and south and it also decreases from the city periphery to rural areas.

The land value in 1971 and in 1981 has similarity in pattern of distribution. In both the periods, it is high near the city limit and it decreases gradually towards the rural areas. In 1971, the highest land value was Rs.65 per square yard in the eastern part of Jodhpur and in 1981 it is Rs.525 per square yard in the north-eastern part of Jodhpur. So, it is interesting to note that the mauja with higher residential growth also has higher land vlaue, though in general one expects to have low demand with high price. This is due to the following factors. Firstly, this land value is the estimated land value of estate agents who always consider Jodhpur as a 'posh' locality. So, they except high land value from this mauja. This perception of a posh area is derived from the high socio-economic status of the original Patel residents of Jodhpur village. Secondly, in Jodhpur, high land value is concentrated in one corner near the city limit where land was already sold to rich communities. Recently, new suburban development is taking place far from the city limit where land value is comparatively lower than that of the city limit. Speculators attract people by creating a 'posh' image of this mauja. They take the opportunity by releasing the plots of larger size in this mauja for richer section who can only afford to pay high price. Gradually it is sold to high middle class people who also have the desire to live in this mauja. They adjust with the high land value by staying far from the city limit and in small houses occupying smaller space.

Vejalpur, in the extreme south, with high residential growth, has an average land value compared to that of Jodhpur. The relatively low land value in Vejalpur is mainly due to the presence of Muslim population in new residential colonies. So the demand of land in this area is limited to the Hindu population of higher socio-economic status. Though people of higher socio-economic status of the Hindu and Muslim religion are now staying together, the land value has not increased very much because this mauja has reached its peak in the

distribution of residential area and there is no scope for further expansion of residential area

Fig 6.22

ISOMETRIC MAP OF LAND VALUE
1981

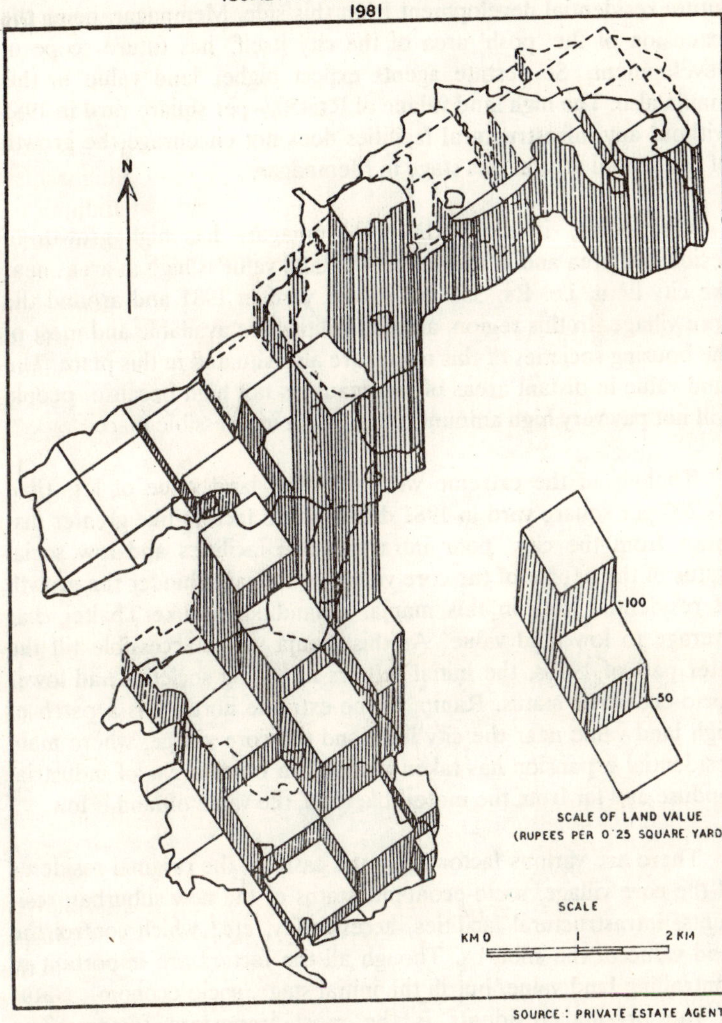

SOURCE : PRIVATE ESTATE AGENTS.

Though Vastrapur and Memnagar, situated in the centre of the study area, do not have very high growth of residential land but speculators perceive bright scope for future expansion of these maujas. It has been depicted by the existence of vast non-agricultural fallow land which will be changed into residential landuse later.Though Vastrapur has a large amount of institutional land, this mauja being the northern extension of Jodhpur, speculators expect future residential development from this side. Memnagar, being the extension of the 'posh' area of the city itself, has future scope of development. So, estate agents expect higher land value in this mauja also. The high land valaue of Rs.450/- per square yard in 1981 without any infrastructural facilities does not encourage the growth of residential area at this stage in Memnagar.

Ghatlodiya, in the north of Memnagar, has high growth of residential area and in this mauja the land value is high in areas near the city limit, i.e. Rs. 350 per square yard in 1981 and around the core village. In this region, all the facilities are available and most of the housing societies of this mauja are also situated in this place. The land value in distant areas of this mauja is not high because people will not pay very high amount to stay in an inaccessible place.

Thaltej, in the extreme west, has low land value of less than Rs.200 per square yard in 1981 due to some factors like greater distance from the city, poor infrastructural facilities and low social status of the people of the core village, which also hinder the growth of residential area in this mauja. Chandlodiya, like Thaltej, has average to low land value. As this mauja was inaccessible till the later part of 1970s, the initial settlers in hosing societies had lower socio-economic status. Ranip, in the extreme north, has a patch of high land value near the city limit and the core village, where main residential expansion has taken place as in other areas of industrial landuse and far from the motorable road, the value of land is low.

There are various factors, like the caste of the original residents of the core village, socio-economic status of the new suburban residents, infrastructural facilities, accessibility, etc., which control the land value of the suburbs. Though all the factors are important in controlling land value, but in the initial stage, socio-economic status of the original residents is the most important factor. The speculators also take this opportunity and they fix higher price in maujas where people of high socio-economic status live. In the later stage, socio-economic status of new residents, factors of accessibility and infrastructural facilities etc. control land value.

TABLE 6.9: **The distribution of population**

A - 1961

	1	2	3	4	5	6
Name of Maujas	Area in sq. Km 1961	Percentage of area	Observed population 1961	% of observed population 1961	Expected population	% of expected lation
1. Vejalpur	7.24	19.14	2704	19.39	2672	19.14
2. Jodhpur	5.11	13.51	478	3.42	1886	13.51
3. Vastrapur	1.13	2.99	430	3.08	417	2.99
4. Memnagar	1.89	5.00	735	5.26	698	5.00
5. Ghatlodiya	4.36	11.52	1006	7.21	1608	11.52
6. Thaltej	8.41	22.24	2224	15.93	3105	22.24
7. Chandlodiya	4.00	10.58	1125	8.06	1477	10.58
8. Ranip	5.68	15.02	5256	37.65	2098	15.02
9. Total	37.82	100	13961	100	13961	100

B - 1981

	1	2	3	4	5	6
Name of Maujas	Area in sq. Km 1981	Percentage of area	Observed population 1981	% of observed population 1981	Expected population	% of expected lation
1. Vejalpur	7.24	19.14	34182	27.51	23783	19.14
2. Jodhpur	5.11	13.51	11739	9.44	16340	13.51
3. Vastrapur	1.13	2.99	4924	3.96	3715	2.99
4. Memnagar	1.89	5.00	6269	5.05	6213	5.00
5. Ghatiodiya	4.36	11.52	15963	12.85	14314	11.52
6. Thaltej	8.41	22.24	6148	4.96	27635	22.34
7. Chandlodiya	4.00	10.52	9675	7.71	13146	10.58
8. Ranip	5.68	15.02	35438	28.52	19112	15.02
9. Total	37.82	100	124258	100	124258	100

Source: District Census Hand Book 1961, 1981.

NB: Chi-square test has been used to calculate distribution of population on the basis of assumption of uniform distribution of population in relation to total area.

2. Demographic Aspects

The pattern of suburban growth is very much associated with its demographic aspects - its distribution, density, growth rate and composition. Like landuse, an analysis of various aspects of demography is also necessary to understand the pattern of suburbs.

(a) *Distribution of population*

In this study area, residential population was concentrated around core villages in 1961. Most of the maujas, except Vejalpur, Memnagar and Ranip, have population less than expected in relation to their total area (Table 6.9A). The expected values of the population have been calculated on the assumption of equal distribution of population over the study area. Two maujas, i.e. Ranip in the extreme north and Jodhpur in the south, has more than 10% variation in the observed (37.65% and 3.42%) number of population than the expected one (15.02% and 13.51%).

In this period of 1961, Ranip was the only urbanised mauja in the study area due to the location of some industries. So, obviously Ranip attracted large number of population to work in these industries in comparison to its area. But Jodhpur was only a part of Vejalpur mauja in 1961 so the population of that mauja was low.

In 1981, two maujas, i.e. Ranip in the extreme north and Thaltej in the extreme west, have 10% variation in observed number of population than the expected one.

Ranip, which was somewhat urbanised in 1961 due to the industrialisation, has been attracted by large number of population at that time. The process of growth of population due to suburbanisation continued in Ranip in 1981. In Thaltej the case is somewhat different. Thaltej has a core village with large number of population in 1961, but it had very low suburban growth due to various physical and social factors, as stated earlier. So, it is obvious that population of Thaltej mauja has decreased relatively in proportion to the total population in the study area in 1981 than in 1961.

So from the chi-square table (Table 6.9), we can accept the

hypothesis 3, i.e. the population distribution is concentrated and not evenly distributed both in 1961 and 1981, at a confidence level of 0.01%. In both cases (1961 and 1981) chi-square test has been used.

The distribution of population in 1961 (Figure 6.23) shows that all maujas except Ranip in the extreme north, had total population of less than 5,000 and in some maujas, i.e. Vastrapur and Jodhpur, it was less than 500 in 1961. In 1981, the base of the population for whole of the study area has increased (430 in 1961 to 4924 in 1981). One mauja, i.e. Vastrapur, has population less than 5,000 in 1981. Two maujas, i.e. Vejalpur in the south and Ranip in the north, have total population of more than 30,000. As a whole, the population has increased in all maujas of the study area but relative growth is high in areas where growth of residential area is also high like Jodhpur and Ghatlodiya.

(b) *Density of population*

The density of persons per unit of area represents the ratio of man to land. Gross density represents the ratio of total population to total land. In 1961, most of the maujas have density of average group (Figure 6.24). While, Ranip in the north has density of very high group and Vastrapur and Jodhpur, in the centre and south, have density of low group. As all maujas, except Ranip, were in the rural stage, so these maujas had density of average to low group. But Ranip, being the only urbanised mauja in 1961, had attracted large number of population which had increased the density of population of this mauja.

The relative situation of gross density has changed over twenty years during 1961 to 1981, due to the suburbanisation, and the value of the density of the average group has also increased from 1961 to 1981 (Figure 6.24). The distribution of gross density in 1981 is very random and it does not show any relationship with the residential landuse (Table 6.8). The relative position of the gross density of some of the maujas, i.e. Vejalpur and Vastrapur, have increased, while in some of the maujas, i.e. Thaltej and Chandlodiya, it has decreased. It is to be noted that maujas whose position has changed to higher group of density are relatively urbanised maujas compared to those maujas whose position of density group have decreased. The development of a mauja in the suburb depends not only on the growth of non-agricultural land but also on the proportionate growth of population.

Fig. 6.23

GROWTH & DISTRIBUTION OF POPULATION
1961 - 1981

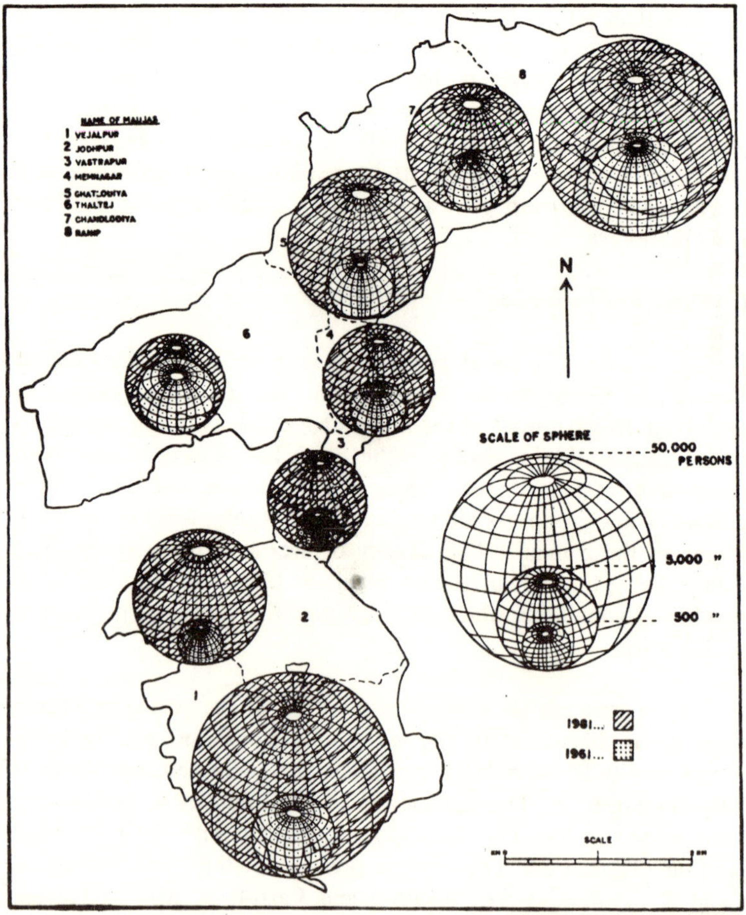

SOURCE : DISTRICT CENSUS HANDBOOK, 1961, 1981.

Net density represents the ratio of residential population to total residential area. As the core villages were the only residential area in 1961, so high density explains the concentration of the large number

Fig. 6.24

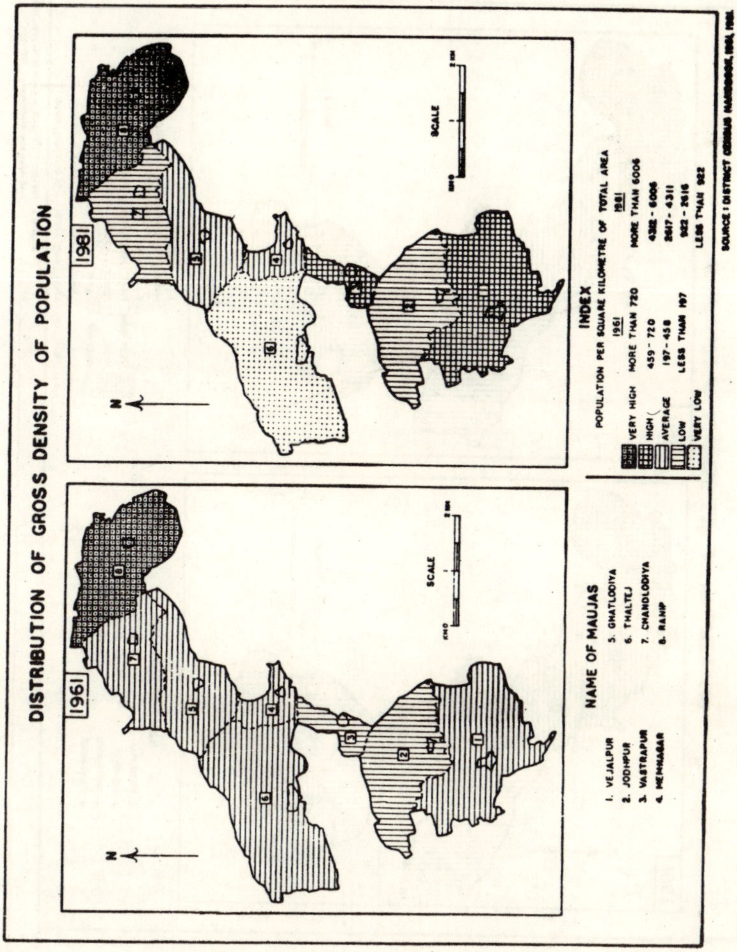
DISTRIBUTION OF GROSS DENSITY OF POPULATION
1961
1981
SCALE
INDEX
POPULATION PER SQUARE KILOMETRE OF TOTAL AREA
1961
1981
VERY HIGH
MORE THAN 720
MORE THAN 5006
HIGH
459 - 720
4312 - 5006
AVERAGE
197 - 458
2617 - 4311
LOW
LESS THAN 197
922 - 2616
VERY LOW
LESS THAN 922
NAME OF MAUJAS
1. VEJALPUR
2. JODHPUR
3. VASTRAPUR
4. MEMNAGAR
5. GHATLODIYA
6. THALTEJ
7. CHANDLODIYA
8. RANIP
SOURCE : DISTRICT CENSUS HANDBOOK

Fig. 6.25

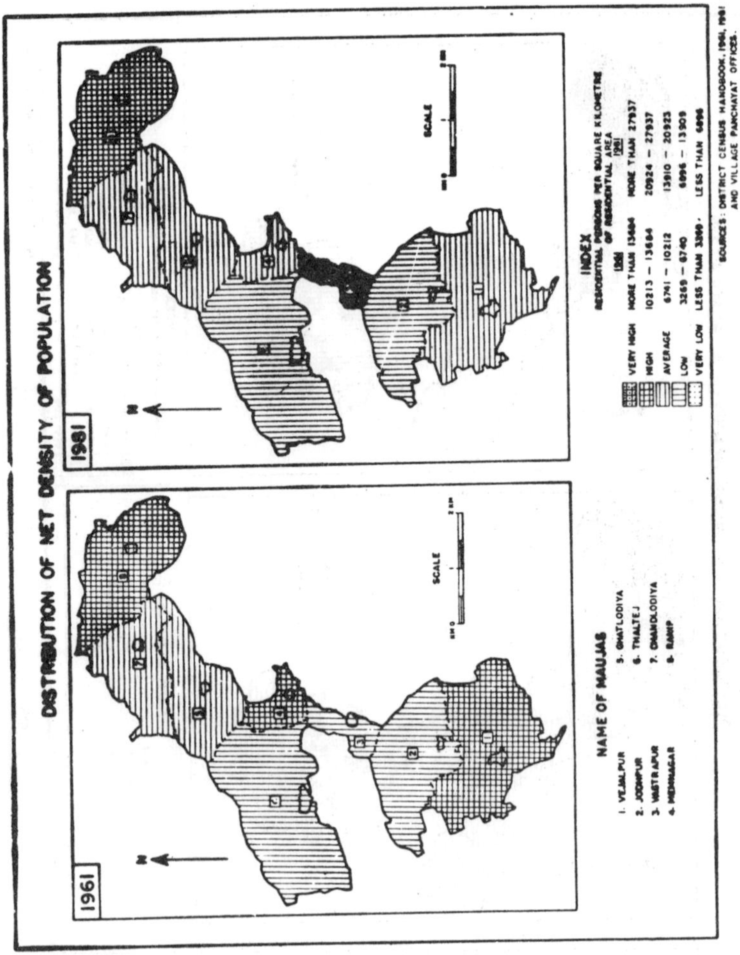
DISTRIBUTION OF NET DENSITY OF POPULATION
1961
1981
N
SCALE
NAME OF MAUJAS
1. VEJALPUR
2. JODHPUR
3. VASTRAPUR
4. MEMNAGAR
5. GHATLODIYA
6. THALTEJ
7. CHANDLODIYA
8. RANIP
INDEX
RESIDENTIAL PERSONS PER SQUARE KILOMETRE OF RESIDENTIAL AREA
1961 1981
VERY HIGH MORE THAN 13684 MORE THAN 27937
HIGH 10213 – 13684 20924 – 27937
AVERAGE 6741 – 10212 13910 – 20923
LOW 3269 – 6740 6896 – 13909
VERY LOW LESS THAN 3269 LESS THAN 6896
SOURCES: DISTRICT CENSUS HANDBOOK, 1961, 1981 AND VILLAGE PANCHAYAT OFFICES.

of population in the core villages. The relative position is more or less same in 1961 to 1981, except in three maujas, i.e. Vejalpur, Vastrapur and Memnagar (Figure 6.25). The higher growth in relative position in density in Vastrapur is due to the existence of high rise buildings in the suburb. Vertical height has not been considered in calculating the residential area.

It is interesting to note that maujas with high residential area have low net density as in Jodhpur and Ghatlodiya. So, it indicates that people in these maujas either live in less crowded manner or they have some open space in their houses, whereas maujas with low percentage of residential area, as in Thaltej and Chandlodiya, net high density indicates congestion in the residential houses.

Growth rate of population (1961-1981)

(c) The growth rate of population is a complex phenomenon, the determinants of which are birth rate minus death rate and inmigration minus outmigration of population. Since none of these elements is available at mauja level, so this study has described the overall growth rate of population with the help of map over time and space.

The study area, as a whole, has experienced an enormous growth of population during 1961-1981 (79%). This high growth of population during the above period in the study area is mainly due to the inmigration of the population in the new suburban areas from the various parts of the city, state and country (Table 5.4 and Figure 5.14).

Most of the maujas have growth rate of population in 1961-1981 from low (less than 44%) to average (1038%) group (Figure 6.26).

The growth rate of population in different maujas follows the same trend as the growth rate of residential area (Figure 6.20). Jodhpur has very high growth rate of population (more than 2032%). Jodhpur, with its residents having high socio-economic status and the core village dominated by Patels, has been an attractive factor to a large number of suburban inmigrants and this inmigration of population has increased the growth rate of population in Jodhpur enormously (Figure 6.26).

Ghatlodiya, in the north, experienced high growth of population (1701%) due to the large number of inmigration in its suburbs in the later half of 70's. Its high growth rate of population matches with the high growth rate of residential area in the same period. It indicates that relative growth of both the residential area and the population in this period (1961-1981) are same, so the relative position of the net density of 1961 and 1981 is same.

Fig. 6.26

DISTRIBUTION OF THE GROWTH RATE OF POPULATION 1961 - 81

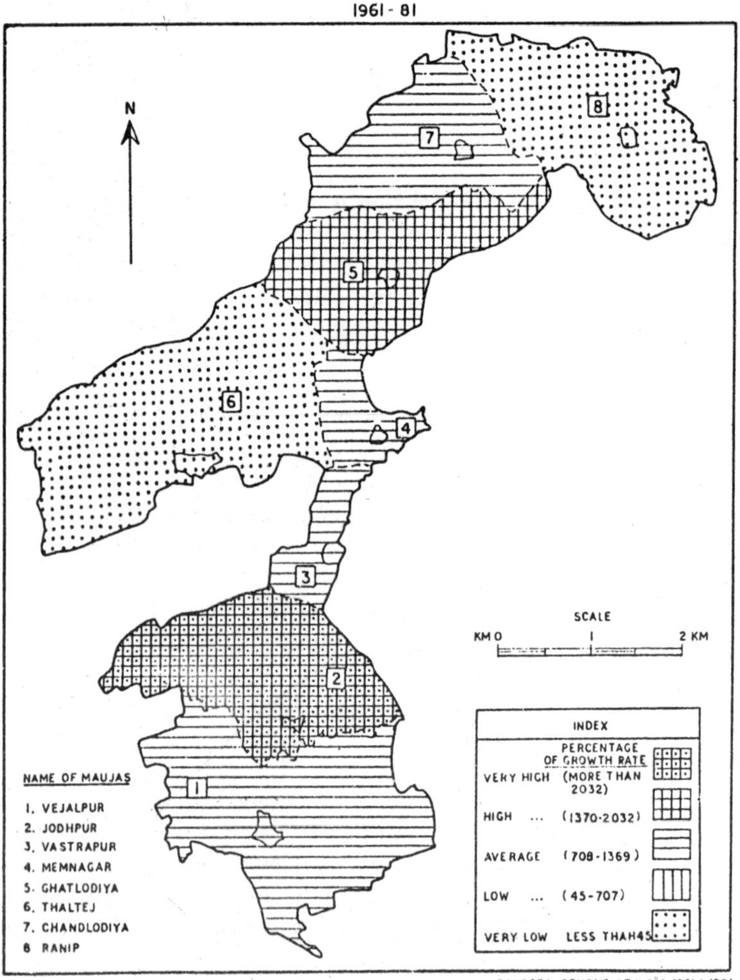

SOURCE: CENSUS OF INDIA 1961 & 1981.

All other maujas, except Ranip in the extreme north and Thaltej in the extreme west, have growth rate of population within average to low group. The growth rate of low group in Ranip is mainly due to two factors, i.e. early suburbanisation has made this mauja congested as early as in 1961, so further growth of population is not very high and, secondly, industrial landuse of this mauja hinders further residential suburban development. In Thaltej, on the other hand, the low rate of suburbanisation is the cause of low population growth rate.

(d) *Composition of population*

Composition of population like sex-ratio, literacy or rate of participation of work indicate indirectly the quality of population. There are other variables like income, caste, mother tongue, age, level of education and occupation which also indicate the composition of population. Due to the non-availability of secondary data regarding various aspects of composition, only sex, literacy and sector of employment of population are considered to study the composition of population in this section. Other variables have been analysed in the next chapter with the help of primary data.

The sex ratio (Female/1000 male) of any area is an efficient demographic indicator, reflecting the population composition of the area and the economic and social implications thereof. In this country, there is a tendency for the urban areas to contain larger number of males and hence, a lower sex ratio is seen due to the sex selective nature of urban inmigration.

The study area being the residential suburb of Ahmedabad has high sex ratio in all decades (Table 6.10). There is only one mauja in the north, i.e. Ranip, which has an industrial zone with 'very low' to 'low' sex ratio in 1961 and 1981, though the absolute number of sex ratio has increased in this mauja from 1961 to 1981. The average sex ratio of the study area has increased as a whole, though in some maujas sex ratio has decreased, the rate of decrease not being very high in any mauja; this variation is mainly due to the natural growth.

Literacy usually reflects the level of achievement of any society. These achievements can be either social or economic. This has been reflected in suburban areas with an increase in the level of literacy in each decade. In all maujas of the study area, the percentage of literate person is increasing (Table 6.11). But this indicator does not measure the level of education totally. This increase in the literate persons may be due to the suburbanisation in which people with higher level of education are migrating in the suburb. On the other hand, due to the change in environment from rural to urban area, the children of the core villages are also attending the primary school. It is not possible to make out the exact cause of lower level of literacy in some maujas, because of lack of data.

TABLE 6.10: **Distributin of sex ratio (female/1000 male) 1961, 1981**

	Name of Maujas	Sex Ratio	
		1961	1981
1.	Vejalpur	875	906
2.	Jodhpur	959	896
3.	Vastrapur	1019	841
4.	Memnger	861	883
5.	Ghatlodiya	961	897
6.	Thaltej	899	853
7.	Chandlodiya	981	879
8.	Ranip	397	824

Source: District Census Handbook, 1961, 1981.

N.B: Due to the presence of large number of bluecollar wrokers in Ranip for its industrial landuse the presence of female population was low in 1961. But it has increased in 1981 due t the increase of migration of bluecollar workers with families and in residental landuse.

The proportion of working population to the total population is an indicator of the economically active component of population. It gives an overall picture of economic dynamism. A high percentage of workers is often associated with a low dependency ratio and vice versa. Percentage of workers in the study area is decreasing (Table 6.12). It is due to the suburbanisation where persons of nuclear families with young dependent children are migrating to the residential suburb.

A higher proportion of non-primary work-force is one of the basic characteristics of the suburban population which means a higher proportion of workers are engaged in activities other than agriculture in urban areas. In the study area, there has been a rapid increase in the percentage of non-primary workers (Figure 6.27). It signifies the diversification of occupation. A cross-sectional analysis of values of this variable for two time periods (1961, 1981) gives an evidence that industrial-cum-residential mauja, i.e. Ranip, has 'very high' to 'high' concentration of non-primary workers at all time (1961, 1981) when compared to the total study area. Two maujas, i.e. Vastrapur and Memnagar, which are very near to the city centre, also have high concentration of non-primary workers.

TABLE 6.11: **Distribution of Literate Persons 1961, 1981 (Percentage to total population**

	Name of Maujus	Literate Persons 1961	Literate Persons 1981
1.	Vejalpur	24	58
2.	Jodhpur	29	66
3.	Vastrapur	17	62
4.	Memnagar	32	67
5.	Ghatlodiya	20	76
6.	Thaltej	25	53
7.	Chandlodiya	30	68
8.	Ranip	36	65

Source: District Census Handbook 1961, 1981.

TABLE 6.12: **Distribution of total workers 1961, 1981 (Percentage of total population)**

	Name of Maujas	Total Workers 1961	Total Workers 1981
1.	Vejalpur	42	27
2.	Jodhpur	40	29
3.	Vastrapur	38	30
4.	Memnagar	40	29
5.	Ghatlodiya	57	28
6.	Thaltej	42	30
7.	Chandlodiya	38	27
8.	Ranip	32	30

Source: District Census Handbook 1961, 1981.

The growth rate of population in 1961-1981 shows very high co-efficient of correlation (0.94) with the growth of residential area and moderate correlation (0.42) with the growth of non-agricultural area during 1961-1981. So, we can accept hypothesis 4 stating that the growth rate of population (1961-1981) is positively correlated to the growth of residential area and non-agricultural area during 1961-1981.

3. Infrastructure

Presence of infrastructure like metalled road, city bus service, educational, medical, recreational and shopping facilities indicate the growth of a suburb. Generally, very small amount of capital is invested for the development of infrastructure in suburbs in India. The settlers of suburbs apparently do not mind being without metalled roads, shops and other amenities, so long as they can have a house. So, population in the suburbs has grown without much development of the area. The individual housing society may have developed its own water supply and sewerge system. The area, as a whole, suffers from lack of amenities, e.g. shopping centre, recreation facilities, fire and police protection, a network of road and trasnport, educational and other institutions, water and sewerage for the whole area.

(a) Accessibility and connectivity

(a) Accessibility and connectivity are important factors which control the growth of a suburb. In this study, accessibility has been analysed with reference to the metalled road (Figure 6.28). Most of the core villages of the study area have relatively 'very high' to 'high' accessibility. In most of the maujas of the study area, there are one or two metalled roads. The accessibility pattern of the study area has become complicated due to the presence of national highway in the western part of the study area. Due to this national highway, some portion of the western part of ths study area has become very highly accessible, but this area being far from the city and not having direct connection with the city-centre, is less suburbanised. Except the western portion of the maujas in the south and eastern part of the maujas in extreme west, zones of 'high' and 'very high' accessibility are concentrated in and around core villages. It is interesting to note that concentration of new housing societies is also in and around the zones of 'very high' and 'high' accessibility. Northernmost mauja, Ranip, has the largest area under zone of 'very low' and 'low' accessibility. Accessibility is good in central and southern portion of the

Fig. 6.27

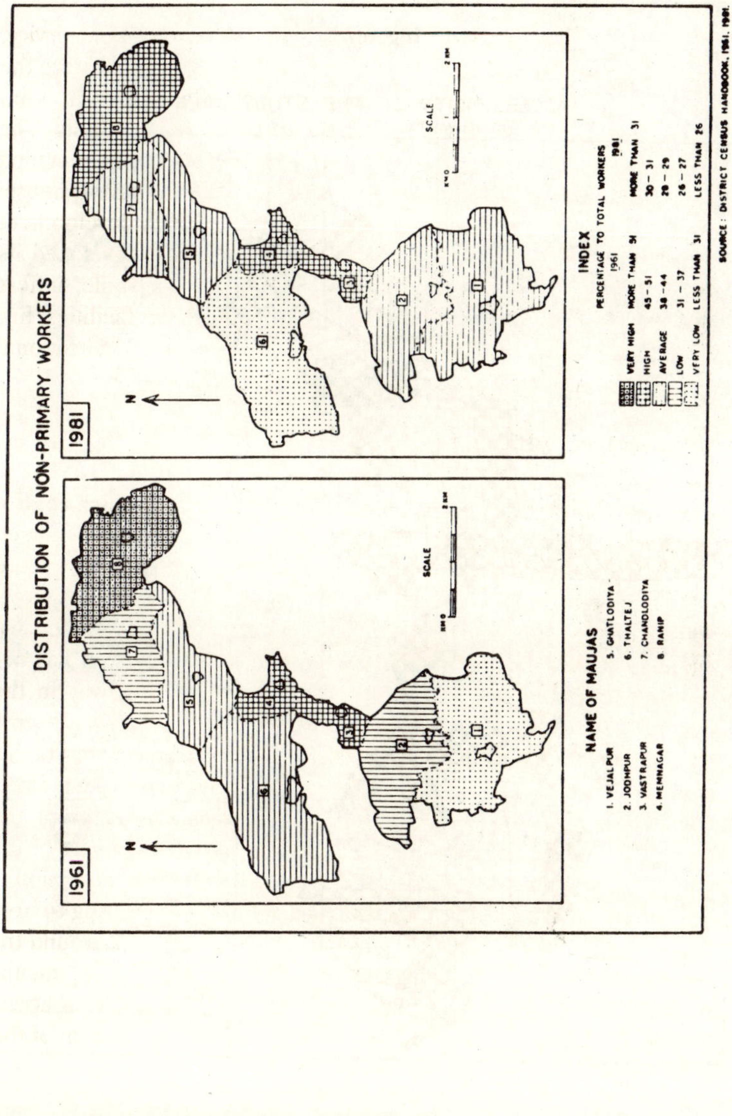
DISTRIBUTION OF NON-PRIMARY WORKERS
1961
1981
N
SCALE
NAME OF MAUJAS
1. VEJALPUR
2. JODHPUR
3. VASTRAPUR
4. MEMNAGAR
5. GHATLODIYA
6. THALTEJ
7. CHANDLODIYA
8. RANIP
INDEX
PERCENTAGE TO TOTAL WORKERS
1961
1981
VERY HIGH MORE THAN 51 MORE THAN 31
HIGH 45 – 51 30 – 31
AVERAGE 38 – 44 28 – 29
LOW 31 – 37 26 – 27
VERY LOW LESS THAN 31 LESS THAN 26
SOURCE : DISTRICT CENSUS HANDBOOK, 1961, 1981.

study area. As a whole, number of metalled road in the study area is not very high and also these available roads are not properly distributed all over the study area. So, accessibility zones follow the same pattern like land value, areas with high accessibility having higher land value than the areas with low accessibility.

Fig. 6.28

ACCESSIBILITY OF THE STUDY AREA
(FROM METALLED ROAD) (1981)

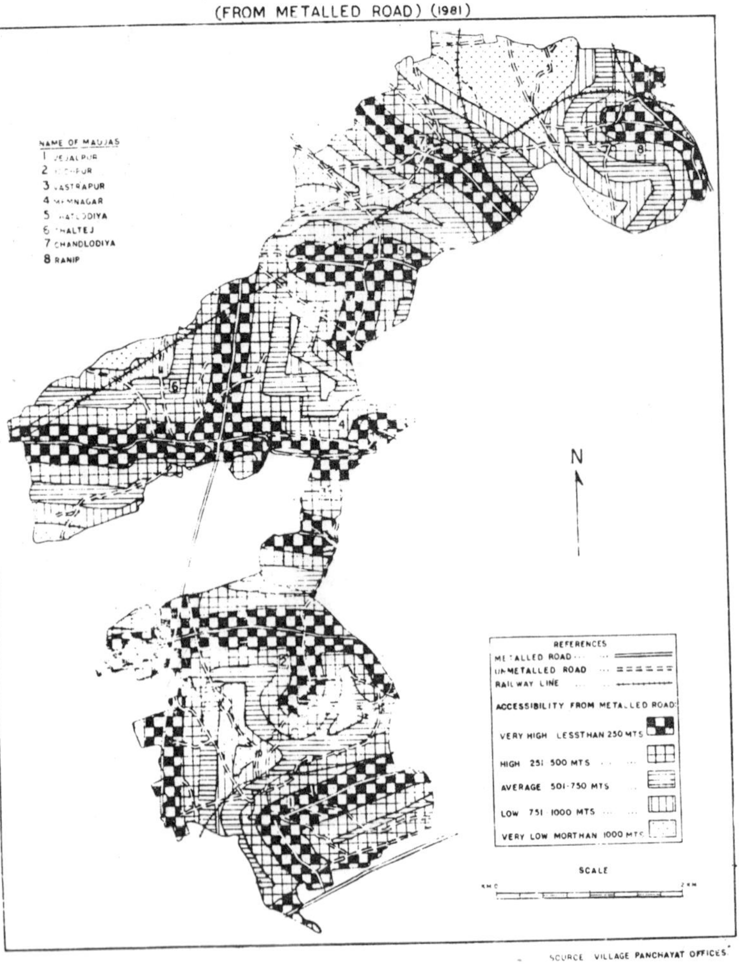

SOURCE VILLAGE PANCHAYAT OFFICES

Connectivity by cheap public transport also controls the growth and development like accessibility. There are 28 bus routes in the study area (Figure 6.39). The bus routes connect suburbs with the city-centre and the main railway station. Though there may be one or two bus routes which are passing through more than one mauja of the study area, there is no bus route connecting the various core vil

Fig. 6.29

CONNECTIVITY ZONE 1981

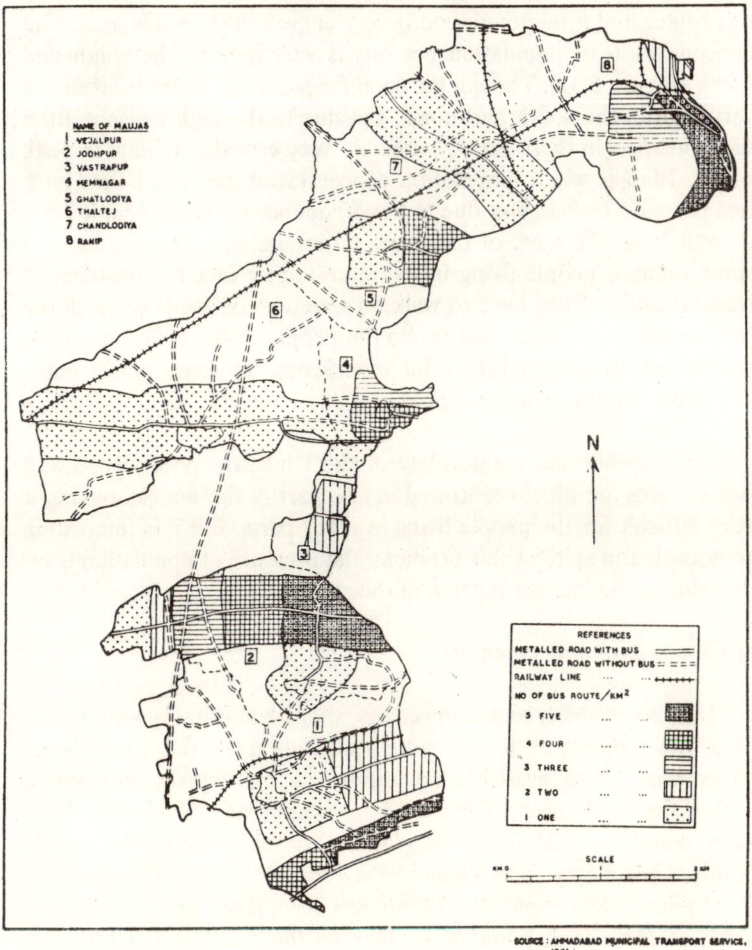

SOURCE : AHMADABAD MUNICIPAL TRANSPORT SERVICE, 1981.

lages. In most of the maujas, there is only one road with bus service. The bus facility is available near the city limit. Southern part of the study area has relatively more bus routes and in this area buses ply in more than one road, serving various parts of the mauja. Concentration of bus routes decreases from south to north-west and again increases in the north. In the northernmost mauja all the buses are concentrated near the city limit.

A large number of bus routes does not indicate high bus frequency (Figure 6.30). The distribution of bus frequency in relation to bus routes and total population is very uneven in the study area. The concentration of population per bus is very high in the south and north (Figure 6.31). Though the total ferquency of buses is relatively very high in the south and north, but due to the high concentration of population in these areas, buses are very crowded mainly in peak hours. Maujas where population concentration per bus is low have less pressure on bus. But due to low frequency of buses, people have to wait long. As most of the buses terminate near the city limit in some maujas, people living in the interior part face the problem of transportation. They have to walk on waterlogged roads to catch the bus mainly in the rainy reason. So the people of the study area have to depend on auto-rickshaw for emergency. In some places auto-rickshaws are not easily available.

Accessibility and connectivity of the study are poor. Road and bus services are all concentrated in one part of the mauja, making it very difficult for the people living in other parts. But it is interesting to note that in spite of this problem, the pressure of population is increasing in the interior portion of the study area.

(b) *Shops and other amenities*

Like accessibility and connectivity, shops and other amenities are very few compared to the total population in the study area. Moreover, these available amenities are not equally distributed within the study area (Table 6.13). Vejalpur and Ghatlodiya have large number of shops in proportion to their population, while Ranip, Chandlodiya and Thaltej have very few shops. Other maujas, i.e. Jodhpur, Vastrapur and Memnagar have more or less the same ratio of shops and population. Due to the scarcity of authorised shops, a large number of unauthorised shops are present in the study area. But most of the shops are located in one place especially at cross-roads in all maujas. The common type of authorised shops in the study area are provision stores, medicine store, groceries, sweet shop, milk shop, tailors shop, saloon, pan shop and flour mill. Other than these authorised shops, there are numerous unauthorised shops

Fig. 6.30

FLOW LINE MAP OF A.M.T.S. BUS SERVICES 1981

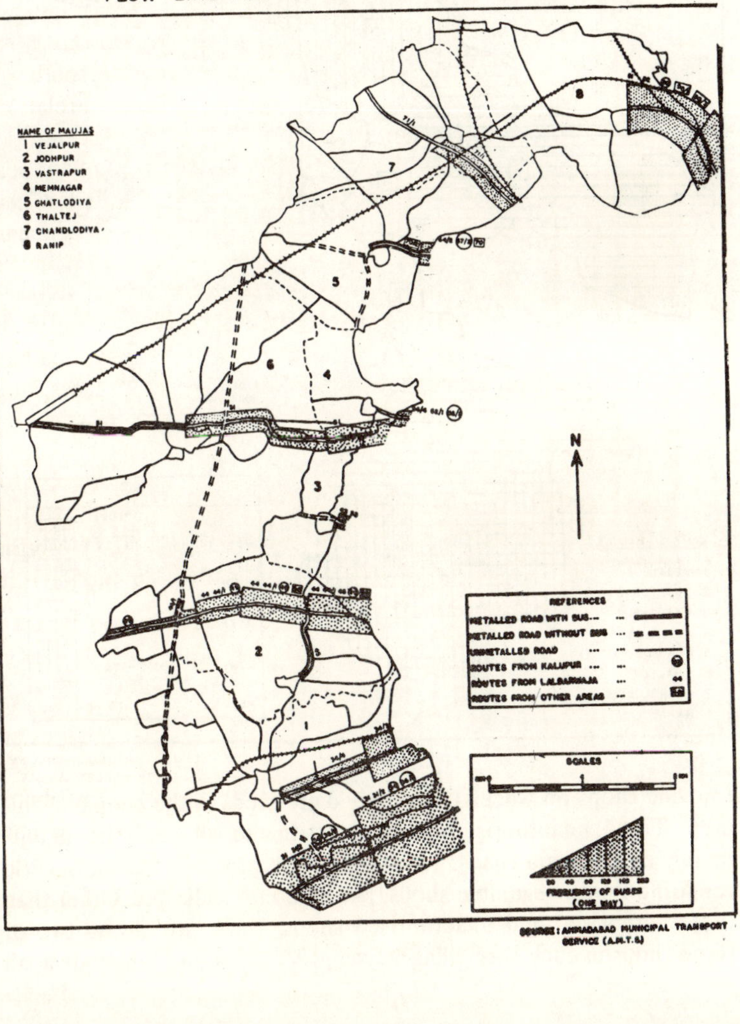

SOURCE : AHMADABAD MUNICIPAL TRANSPORT SERVICE (A.M.T.S)

Fig. 6.31

POPULATION CONCENTRATION IN THE PUBLIC TRANSPORT 1981

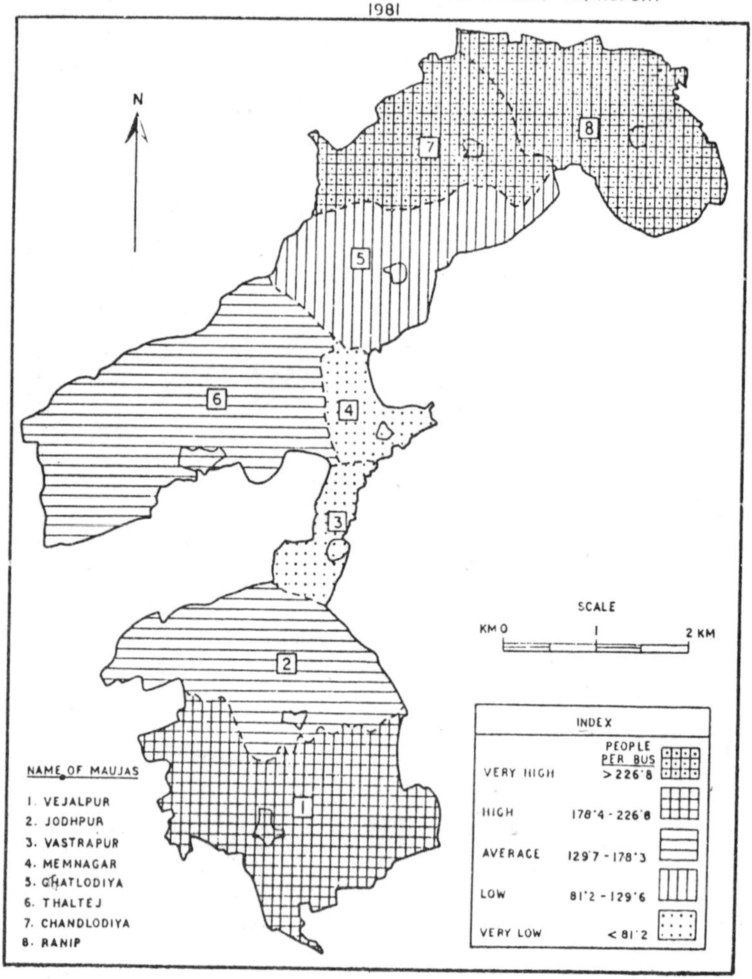

SOURCE: AHMADABAD MUNICIPAL TRANSPORT SERVICE.

(moving shops on wheels) to supply daily needs to the people of this area. These unauthorised shops are situated all over the mauja, mainly on the main roads. These shops are generally tea stalls, cycle repairing shops, vegetable shops, pan cigarette stalls, etc. Other than these authorised and unauthorised shops, there are always two or three shops in each core village. Except Vejalpur and Ghatlodiya, all maujas of the study area have insufficient shopping facility. Jodhpur has six 'posh' shopping centres, situated in various parts of the mauja. Prices of commodities are high in these shopping centres, compared to those at the core city.

TABLE 6.13: **Distribution of authorised shops (1981)**

		1	2	3	4	5	6
	Name of Maujas	Observad population	% of observed population	Observed of shops	% of observed shops	Expected of shops	% of expected shops
1.	Vejalpur	34182	27.51	223	40.40	152	27.51
2.	Jodhpur	11739	9.44	54	9.78	52	9.44
3.	Vestrapur	4924	3.96	18	3.26	22	3.96
4.	Memnagar	6269	5.05	38	6.88	28	5.05
5.	Ghatlodiya	15963	12.85	152	27.54	71	12.85
6.	Thaltej	6168	4.96	8	1.45	27	4.96
7.	Chandlodiya	9575	7.72	12	2.17	43	7.71
8.	Ranip	35438	28.52	47	8.52	157	28.52
	Total	124258	100	552	100	552	100

Source: District Census Handbook 1981 and field survey by the author.

Note: Chi-square test has been used to calculate expected frequency of authorised shops on the basis of assumption of uniform distribution of population.

Therefore, suburban population depends mainly on the city-centre for shopping. In Vejalpur and Ghatlodiya, most of the people depend on local shops for buying daily food and household items, because the number of shops are plenty so prices are reasonable due to competition. In Ranip and Chandlodiya, though number of shops are very few and price is also high compared to that of city-centre, people still prefer to buy daily food and household items from the local shops rather than spending money and time in transport to go to the city-centre. People living in Jodhpur, Vastrapur and Memnagar depend on the city-centre for daily food and household items. Regarding clothes and other costly household things, suburban people depend on the city-centre.

Like shopping facilities, suburban people mostly depend on city core for jobs, service of police, post-office, bank, education, medical

Fig. 6.32

DISTRIBUTION OF AMENITIES
1981

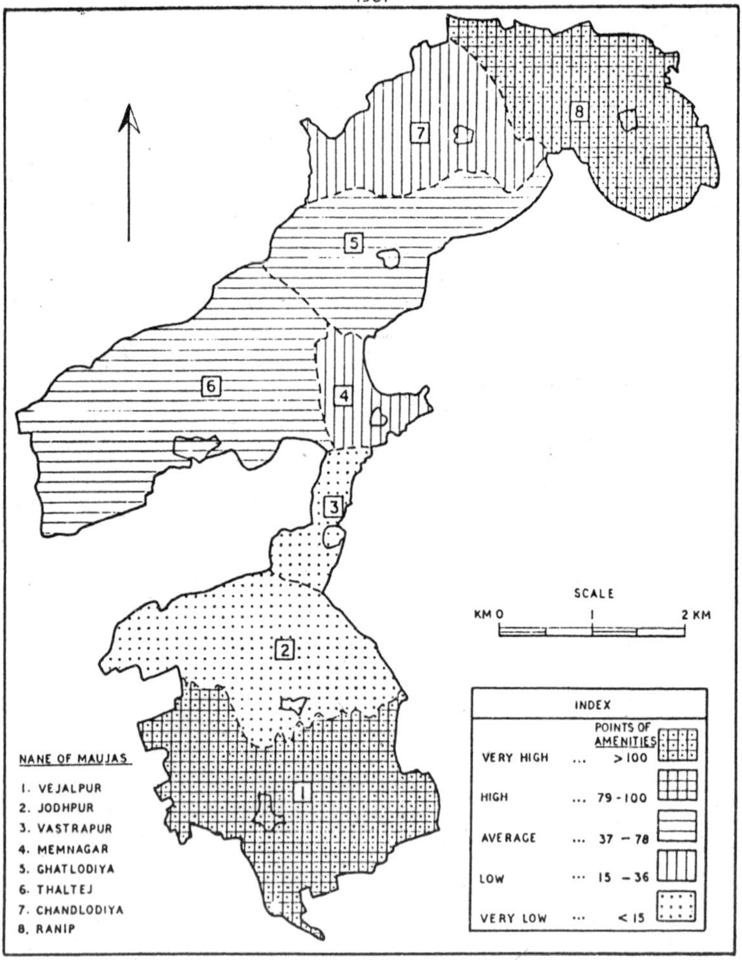

SOURCE: AHMADABAD URBAN DEVELOPMENT AUTHORITY.

facilities, entertainment and religious festivals. It is interesting to note that the growth of residential area does not have any relationship with the available facility in any mauja (Figure 6.32). Maujas with high amenities are situated in extreme north and south. Maujas with average amenities are scattered all over the study area. Many maujas of the study area have low to very low amenities. Distribution of amenities also does not have any relation with the growth and development of population. Therefore, it can be concluded that growth and development of Indian suburbs are not associated with the availability of various shopping and other facilities.

As the available facilities are not equally distributed over the study area in relation to population, hypothesis 5 that the frequency of buses, shops and other amenities are equally distributed in all maujas in relation to total population is rejected by using chi-square test at a confidence level of 0.01%.

4. Summary of the Chapter

Landuse of the study area has gone through an enormous process of conversion. This conversion of agricultural land into non-agricultural landuse has been controlled by speculators who own most of the non-agricultural, vacant land of the study area. The share of agricultural land has decreased gradually with the increase of various non-agricultural land, especially residential area. An increase in non-agricultural fallow indicates the scope of future urban expansion, as non-agricultural fallow gets converted into residential area. First, fallow land is seen near the city limit and then it has spread along the roads and it has encroached interior land. The growth of land value of residential land in the study area is controlled by physical and social factors, e.g. distance from city, accessibility, demand for the land, socio-economic status of the people of core villages. The distribution of residential land and land value are very much correlated in a positive manner. In other words, the area which is developed in residential landuse has higher land value compared to those areas which are fallow lands and are yet to develop.

Like landuse, demographic aspects of the study area have also changed considerably. There is an increase in both gross (persons per sq.km) and net (residential persons per sq.km of the residential area) density. Relative position of gross density of some of the maujas has also changed from 1961 to 1981. It is interesting to note that maujas which are more urbanised, e.g. Vejalpur, has changed from average to higher group of density of population, while less urbanised mauja, e.g. Thaltej, has changed from average to lower group of gross density. But in case of net density, maujas with high residential area, e.g. Jodhpur, have low density which indicates that people in these maujas live in less crowded manner, having houses with more space per person. The study area as a whole has enormous growth of total population and of residential area. The influence of suburbanisation in the study area is felt with the increase of sex ratio, literacy and percentage of non-primary workers.

The study area has few metalled roads, city buses, authorised shops and other amenities. These few infrastructural facilities are not equally distributed all over the study area. Southern (Vejalpur and Jodhpur) maujas and central (Vastrapur and Memnagar) portions of the study area have comparatively good accessibility, i.e. more areas are nearer to the metalled roads. Connectivity is also high in south (Vejalpur) and north (Ranip). Maujas situated in south, north and north-west have high to very high concentration of commuters per bus. In most of the maujas, the bus service is concentrated in one road and it is very difficult for the people living far from the road to catch the bus. Authorised shops are large in number only in Vejalpur and Ghatlodiya. Except in Jodhpur, in all other maujas shops are concentrated on the main roads. Other amenities are also not equally distributed over the study area in relation to the distribution of population. The growth of population has low relationship with the availability of infrastructural facilities.

SEVEN

THE SOCIO-ECONOMIC STATUS OF RESIDENTS

The socio-economic status of residents of the core villages and the housing societies have been analysed with the help of eighteen variables (Appendix 4 A-D). In these tables, detailed data have been tabulated by each mauja and with sub-divisions of core villages and housing societies for individual variable. These eighteen variables have been grouped into four sub-factors, i.e. information regarding the head of the family: general information regarding family; female participation in the family and socio-cultural-economic background of the total family. All data used in this analysis have been collected by the field survey. In this chapter, analysis has been done first by comparing between residents of all core villages and of all housing societies, then between core villages and housing societies of each mauja. These comparisons will be done with the help of 'Modernity Index'. Method of calculation of 'Modernity Index' has been discussed in section 2 of Chapter-III. If the score in 'Modernity Index' is more than 100, in a mauja for a certain factor, that factor will be considered modern in that area; otherwise, the factor will be traditional if the score is less than 100.

The residents of core villages due to their rural origin, age-old outlook, belief and customs are traditional in nature; so they have low score in 'Modernity Index' than the residents of the housing societies in all four sub-factors of the socio-economic status. But, it is also true that there are variations between the core villages and between housing societies of various maujas. It is not that residents of all housing societies are modern and all of core villages are very traditional.

1. Information regarding head of the family

(a) *All core villages versus all housing societies*

In the core villages, most of the people are living in their places of birth, i.e. the core village itself and have low level of education, that is 18% of them are illiterate and 68% have only primary level of education. The residents (73%) of the core villges also practise traditional occupation of their own caste, like potter, milkman, cultivator, etc. In housing societies most of the people are inmigrants (98%). Most of them have come from various parts of Gujarat and India (84%). It is interesting to note that the percentage of population from the core-city to the suburbs is small compared to that of inmigrated population from other parts of Gujarat and India. It is mainly because the majority of the people from the core-city have outmigrated and have settled in other parts of the city and some of them only have shifted to the suburbs. Inmigrated people also first stayed within the city for few years mainly in rented houses and then a large number of them hve shifted to the suburbs making their own houses or getting house in cheaper rent. The heads of families living in housing societies have higher level of education, i.e. only 18% have educated upto primary level, whereas 20% have education more than the graduate level. They are also engaged more in non-primary sector of occupation (83%). Due to their low level of occupation and education, the residents of the core villages get low score in 'Modernity Index' than the residents of housing societies (Figure 7.33).

(b) *Analysis between core villages*

There is also some difference between the residents of different core villages regarding their level of education and occupation. Vastrapur, which is an institutional-cum-residential mauja, has a large number of residents (52%) of the core village engaged in various non-primary activities. They are working mainly as peon, sweeper and gardeners in these institutions. Other than original core villagers, there are people (32%) from different parts of Gujarat and

Fig. 7.33

INFORMATION REGARDING THE HEADS OF THE FAMILY 1981

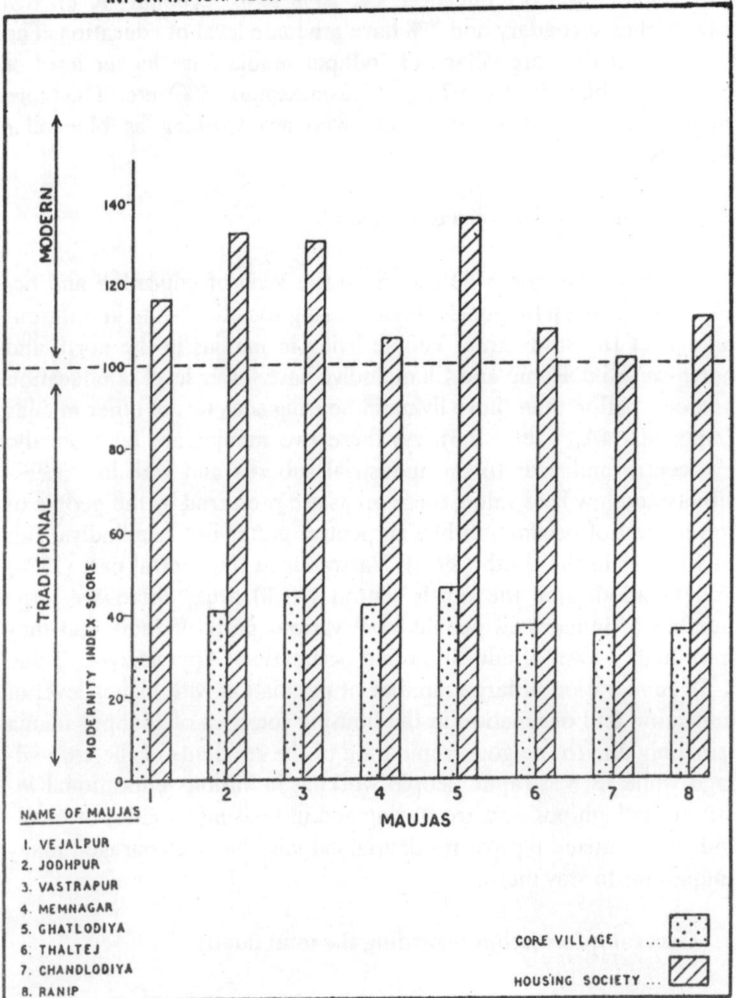

SOURCE: FIELD SURVEY BY THE AUTHOR.

India, who also work in these institutions and live in cheap rented houses which are available in core villages. The situation is same in Ranip, an industrial-cum-residential suburb in the north. In Ranip also a large number of people (40%) are engaged in non-primary activities, mainly as unskilled workers and they stay at the core village. But the situation is quite different in the core village of Jodhpur, a true residential suburb of Ahmedabad. In this core village, people have higher level of education, i.e. 20% have secondary level. 6% have higher secondary and 7% have graduate level of education. The residents of the core village of Jodhpur mauja have higher level of occupation like clerk (13%) and businessman (8%) etc. This core village have very few inmigrants who are working as bluecollar workers.

(c) *Analysis between housing societies*

There is also some difference in the level of education and occupation between the residents of housing societies living in different maujas of the study area. People living in maujas in the north and north-west, i.e. Ranip and Chandlodiya have lower level of education and occupation than those living in housing societies of other maujas (Appendix 4A, Table 2,3). As these two maujas are far from the city-centre and near to the industrial suburbs and with low accessibility and low land value, these areas are preferred by the people of lower level of occupation like carpenter, potter in Chandlodiya. But in Jodhpur in the south (29%), Vastrapur in the central part (27%) and Ghatlodiya in the north central (20%) area, there are large number of inmigrated people from various parts of India and they have higher level of education and occupation (Appendix 4A, Table 1-3). Inmigration of large number of population with higher level of education and occupation in the housing societies of Jodhpur mauja is mainly due to the good reputation of the residents of the core village while in Vastrapur people working in various educational institutes in high position are staying and in housing societies of Ghatlodiya, the mixed type of residential colonies have encouraged many inmigrants to stay there.

2. General information regarding the total family

(a) *All core villages varsus all housing societies*

As the rural culture is still existing in the core villages, the joint family system (64%) and large number of family members, i.e. more than 6 persons (63%) (indicators of traditional life-style) are still very common there. Due to the joint family system, grandchildren and great-grandchildren are also very common in a family of the core villages. But not only direct relations but distant relations are also staying in the same house in the core villages (Appendix 4B, Table 6). Early marriage (below the age of 20), another indicator of traditional life-style, is common (29%) among both males and females in all core villages. Though younger generation (62%) has started to go to schools, their social environment does not allow them in taking up higher education, i.e. more than school level. So, very few of them (10%) are engaged in various non-primary, high level of occupation. Families in the housing societies are small, i.e. most (67%) of them have less than six persons in a family and are of nuclear (58%) type, where husband, wife and two or three unmarried children are common. But due to the traditional nature of the Indian society, joint family system is also common in housing societies, where married son, daughter-in-law and grandchildren or widowed mother are living in the same family. Early marriage is not very common (4%) within the people of housing societies. Most of the residents of housing societies have higher level of education and due to the concentration of large number of younger residents, many (45%) of them are students (Appendix 4B, Table 9,10). So, it is obvious that residents of core villages have low score in the 'Modernity Index' than the residents of housing societies (Figure 7.34).

(b) *Analysis between core villages*

There are few dissimilarities between residents of core villages of different maujas regarding the socio-economic aspects of their family. In the core villages of Vastrapur (68%) and Ranip (60%), large number of inmigrants are living and there relatively few joint families are seen. Like the head of the family of the core villages of Vastrapur and Ranip, other family member are also engaged in various non-primary activities of low skill while in the core villages of Jodhpur, family members have comparatively higher livel of education and occupation.

(c) *Analysis between housing societies*

Residents of housing societies as a whole have high score in the

'Modernity Index' regarding the socio-economic characteristics of

Fig. 7.34

INFORMATION REGARDING THE FAMILY 1981

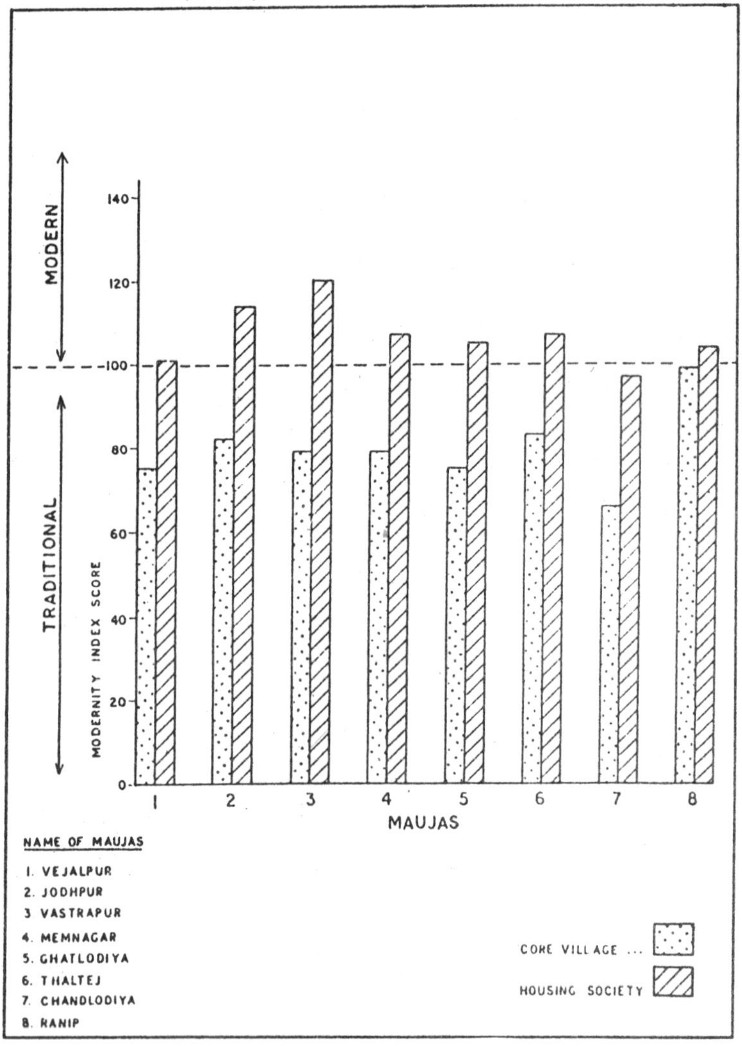

family members. But residents of housing societies of Ranip (9%) have traditional style of early marriage. In Jodhpur and Vastrapur, there are no married male and female of early age, that is below the age of 20, in housing societies. Very few houses in Jodhpur (9%) have traditional type of large number of family member while in Vejalpur (68%) many households have large number of family members. Regarding other variables, i.e. family system, number of children, level of education and occupation, there are minimum variations between households living in the housing societies of different maujas.

3. Female participation in a family

(a) *All core villages versus all housing societies*

Traditional nature of the low female participation of the residents of core villages has been reflected in the low level of female literacy (42%) and low level of occupation (100% female of core villages are housewives) (Figure 7.35). The number of family female students in the age group of 5-20 is also very low (53%) in the core villages. The people living in housing societies have more score in 'Modernity Index' than the people of core villages regarding different aspects of female participation (Figure 7.35). Females of housing societies are literate (88%); most (94%) of the female children in the age group of 5-20 are students and females (14%) in housing societies are engaged in various jobs, other than household works.

(b) *Analysis between core villages*

The residents staying in the core village of Jodhpur have relatively high (78%) female literacy and 73% of female students at the age group of 5-20 years, whereas in Chandlodiya and Ranip, percentage of female literacy (50% and 42% respectively) and female students (43% and 48%) are comparatively lower than that of the core villages of other maujas. Variations in female participation within various core villages depend on the cultural and social background of the people of that village (Appendix 4C, Table 11-13).

(c) *Analysis between housing societies*

People living in the housing societies of Jodhpur mauja have relatively high score in 'Modernity Index' than the residents living in housing societies of other maujas. In the housing societies of

Fig. 7.35

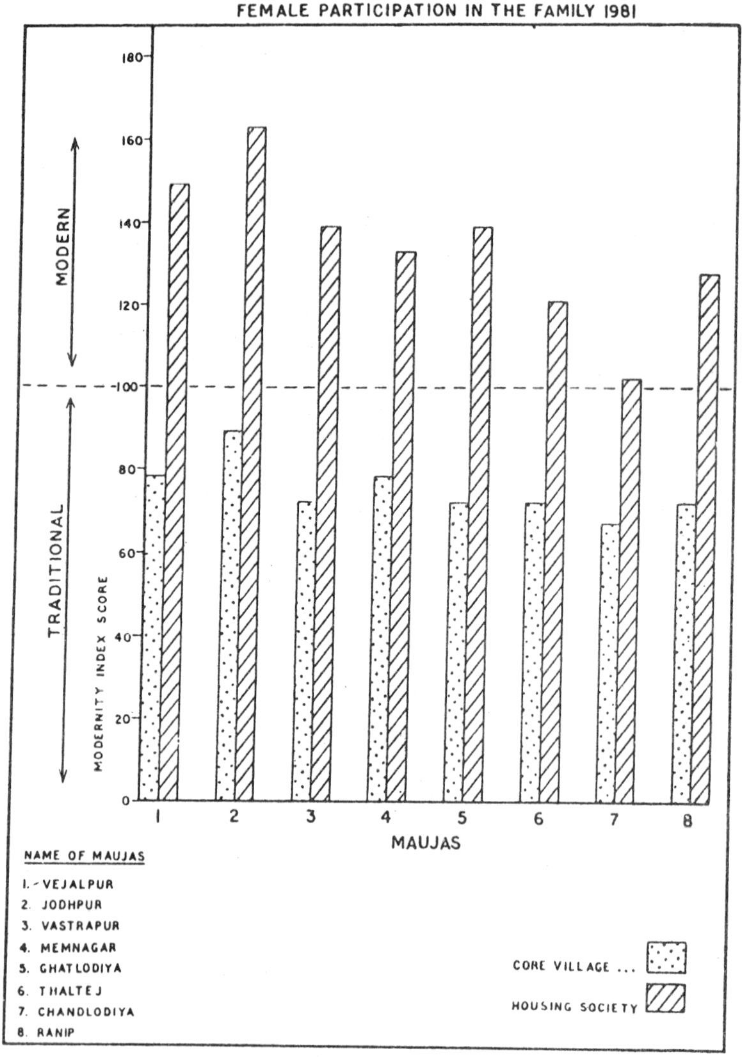

Jodhpur and Vastrapur maujas, percentage of female student (100%) and female literacy (100%) are the highest in the study area. In the housing societies of Jodhpur, many females (25%) are engaged in services and some of the(12%) are engaged in managerial post and professional work. In Vejalpur, situated in the extreme south, also a large number (24%) of female population of the housing societies of this maujas is engaged in services, including in managerial post and professional work. But in Ranip and Chandlodiya, the residents of housing societies have relatively low score in 'Modernity Index' because female literacy (81% and 80%) is comparatively lower than that of the housing societies of other maujas; so, none of the female population living in housing societies of these maujas are engaged in any outside jobs.

4. Information regarding socio-cultural-economic background of the family

(a) *All core villages versus all housing societies*

The socio-cultural-economic background of a family includes residents' food habit, religion, mother tongue, caste and total monthly income of a family (Appendix 4D, Table 14-18). Core villages have been considered traditional because, most of its residents having traditional food habits, i.e. vegetarian (93%), Hindu religion (93%), Gujarati as mother tongue (95%) and low total monthly income of the family, i.e. less than 1000 rupees (80%). In housing societies, as most of the people have migrated from various parts of Gujarat, they are Gujarati speaking (84%), Hindu (87%) and vegetarian (72%). But very few of them (4%) have very low total family monthly income, i.e. less than 500 rupees and some of the (10%) have very high total family income, i.e. more than 2000 rupees. In general, residents are more cosmopolitan in housing societies. There are some housing societies based on religion though. As a whole, residents of housing societies have high score in 'Modernity Index' regarding socio-cultural-economic background of the family than the residents of core villages (Figure 7.36).

(b) *Analysis between core villages*

There are some variations regarding the socio-cultural-economic background of the family between different core villages of the study area. In core villages of Ranip and Vastrapur, there are people from various parts of Gujarat and adjacent states, i.e. Rajasthan and

Maharashtra (Appendix 4D, Table 16). These inmigrated population

Fig. 7.36

INFORMATION REGARDING SOCIO CULTURAL ECONOMIC BACKGROUND OF THE FAMILY
1981

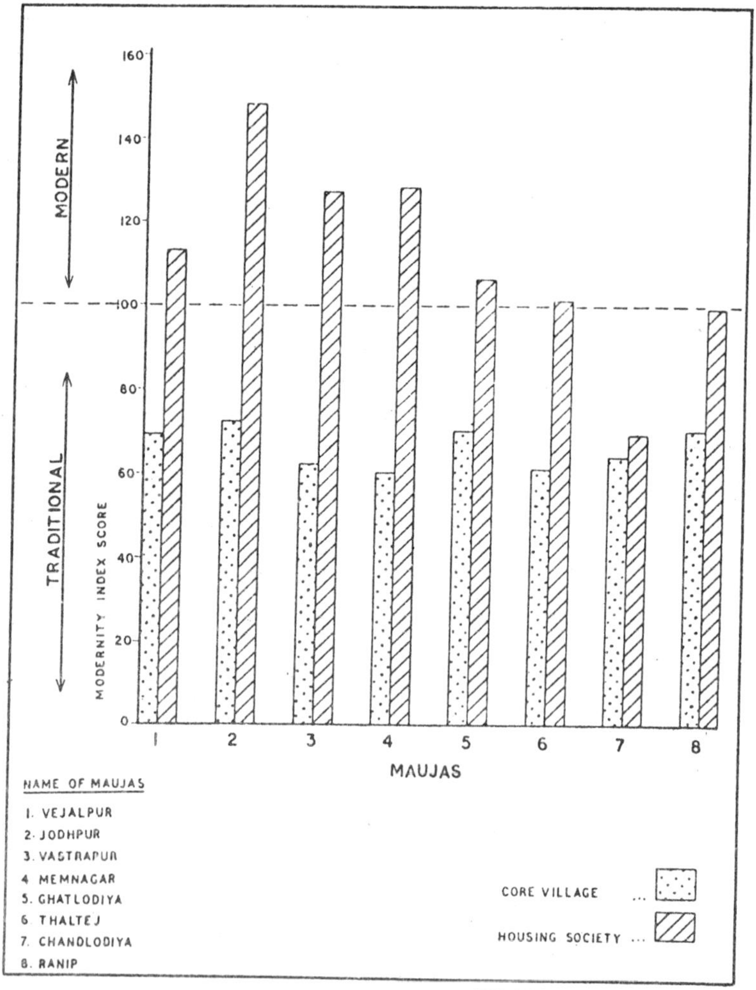

SOURCE: FIELD SURVEY BY THE AUTHOR.

working in various institutions and industries situated in these maujas, have increased the percentage of non-Gujarati, non-Hindu and non-vegetarian population in the core villages of the maujas. Most of these core villages, except Ranip, are dominated by one caste (Appendix 4D, Table 17). Though in Vastrapur the people from other areas have also inmigrated, they are not much dominating in the core villages. There is a sharp contrast between the residents of core villages regarding their total monthly income of the family. All families of core villages of Ranip and Chandlodiya have low monthly income, but in core villages of Jodhpur residents are well-to-do (66% of households have low income and 14% have high income i.e. more than Rs. 1500, of which 7% have very high income i.e. more than Rs. 2000).

(c) *Analysis between housing societies*

In the maujas situated in the central and southern part of the study area, i.e. Memnagar (30%), Vastrapur (21%) and Jodhpur (25%) have more number of non-Gujarati speaking people in the housing societies. This is mainly due to the presence of various national organisations and institutions where employees are from other states of India. Vajalpur in the extreme south has the highest concentration (20%) of Muslim population in its housing societies. In this mauja, some Muslims had migrated from the core city of Ahmedabad in the later part of 1960's and other Muslim population from the core city had followed them gradually after finding that Vajalpur is safe for them to live. So they have formed religion-oriented societies in the southern part of Vejalpur. There is also one housing society of Christian community each in Vejalpur and Jodhpur and one Tamil concentrated housing society is found in Jodhpur. But, in general, Jain, a non-Hindu religion, live along with Hindus in housing societies. Regarding the variation of total family income of households between housing societies, Chandlodiya has the highest (79%) number of households with low income and none of the households have high income, while in Jodhpur the percentage of households with low income is 27% and with high income is 47% (Appendix 4D, Table 18). This variation in income is mainly due to the variation in the level of education and occupation between the residents of housing societies of various maujas.

So, we can accept the hypothesis 6, i.e. there is difference between residents of core villages and housing societies regarding their socio-economic status at .05% level of confidence by using Analysis of Variance test and can confidently assert that the residents of housing societies are more "modern" than those of core villages from the point of view of socio-economic status.

5. Summary of the Chapter

Residents of core villages due to their rural origin and belief in age-old customs are more traditional. This has been indicated by the low level of education, traditional hereditary occupation, joint family system, large number of family members, low female participation in work, vegetarian food habit and low average monthly income due to low level of occupational skill. On the other hand, residents of the housing societies being educated, have modern occupation and small size of family which indicate that they have modern socio-economic status. There is also much variations between core villages and between housing societies situated in various maujas. As a whole, maujas situated in the north and north-west, i.e. Ranip and Chandlodiya, have residents of low socio-economic status, both in their core villages and in housing societies. This is mainly due to the concentration of bluecollar workers in Ranip; and poor location of Chandlodiya has not attracted residents of high socio-economic background. On the other hand, residents of both the core village and housing societies of Jodhpur Mauja, situated in the south, have high socio-economic status. This is mainly due to the attraction of original residents of the core village who are Patels and have high socio-economic status in the society in Gujarat. So, the newly people settled people are also with high socio-economic status,being attracted by the Patels of the core village.

EIGHT

BEHAVIOUR AND LIFE STYLE, PERCEPTION, ATTACHMENT AND INTERACTION PATTERN OF SUBURBAN POPULATION

1. Behaviour and life style

Behaviour and life style of suburban population have been analysed with the help of twentysix variables (Appendix 4E-I). these twentysix variables have been grouped into five sub-factors, i.e. possession and use of modern goods; reading and recreation habits; the behaviour in raising children; attitude towards banks and saving; and knowledge of city and suburbs. All these data used in this analysis have been collected by field survey. In this Chapter, analysis have been done, first by comparing between residents of all core villages and of all housing societies, then between core villages and housing societies of each mauja. These comparisons have been done with the help of 'Modernity Index' similar to that of socio-economic status in the previous Chapter.

Like socio-economic status, residents of core villages have traditional outlook regarding various aspects of life style, so they have low score in 'Modernity Index' than the residents of the housing societies in all five sub-factors of the behaviour and life style. Low scores in 'Modernity Index' of residents of core villages in some factors, i.e. reading and recreation habits and possessioin and use of modern goods, are due to the lower level of income of the residents of core villages. But the low scores of some variables (i.e. first preference given to caste in children's marriage, expectation of lower level of education for female child than male child etc.) indicate that in the housing societies housing characteristics have changed faster than that of behaviour of residents.

(a) *Possession and use of modern goods*

(i) *All core villages versus all housing societies*

The residents of core villages possess and use very few modern goods. None of them possess record player, television, dining table, refrigerator, telephone, etc., and very few of them possess pressure cooker (2%) and gas stove (5%), few of them use soap for cleaning utensiles (6%) while most of the residents of housing societies have pressure cooker (76%), gas stove (54%) and some of them have dining table (30%) and refrigerator (30)%) and most of them use soap for cleaning utensils (94%) (Appendix 4E, Table 19, 20). So, obviously residents of housing societies have high score in the 'Modernity Index' than the residents of core villages in this factor (Figure 8.37).

(ii) *Analysis between core villages*

The residents of the core villages of Jodhpur possess and use modern goods. It is also true for the residents of the core village of Ghatlodiya. As the residents of the core village of Jodhpur have higher income, they can also afford to possess and use modern goods like gas stove (27%), radio or transistor (80%), scooter (20%), etc. Same in the case for Ghatlodiya and Vastrapur where people are also engaged in secondary and tertiary jobs and try to change their life style by possessing and using modern goods. In some maujas, were people have non-primary occupation (like Ranip), they do not possess and use modern goods. This is due to lower level of income and also due to their rural life style. No residents of the core village of Ranip possess tape recorder, scooter, etc.

(iii) *Analysis between housing societies*

People of the housing societies of Jodhpur with higher level of income and with urban life style possess and use modern goods. So, they have high score in 'Modernity Index'. They are followed by the

Fig. 8.37
POSSESSION AND USE OF MODERN GOODS
1981

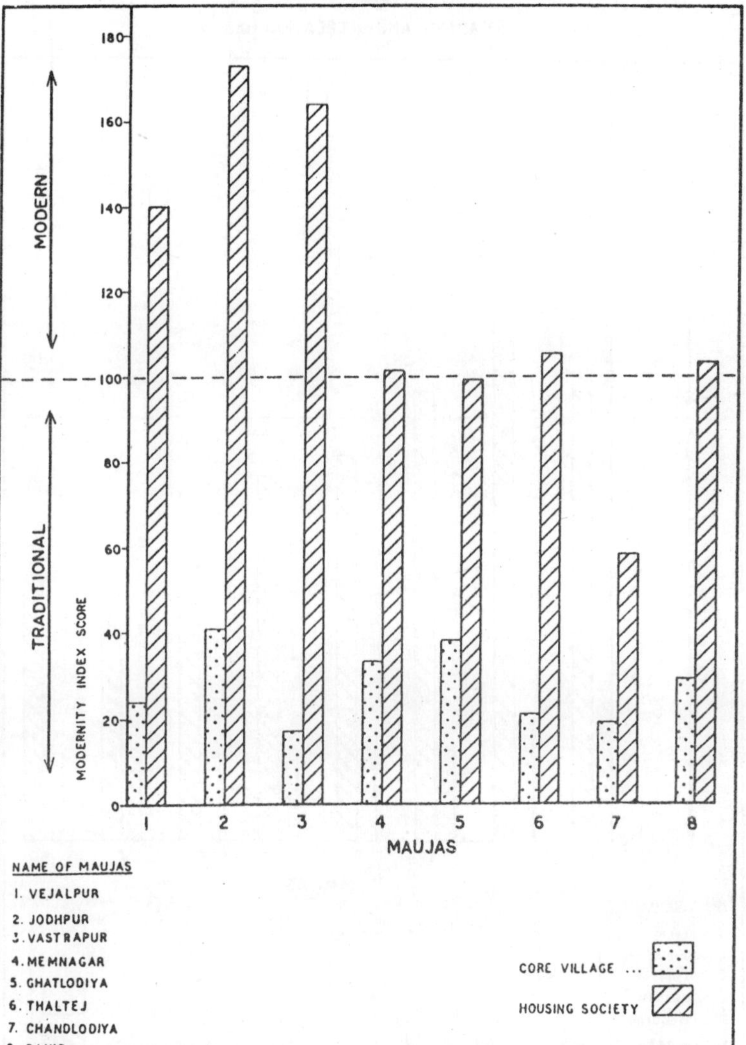

people of Vastrapur, who also have high income and higher level of occupation. Residents of Ghatlodiya have higher level of occupation but they do not possess or use the modern goods. The residents of Chandlodiya and Ranip, due to their lower level of income, do not possess or use modern goods.

Fig. 8.38

READING AND RECREATION HABITS
1981

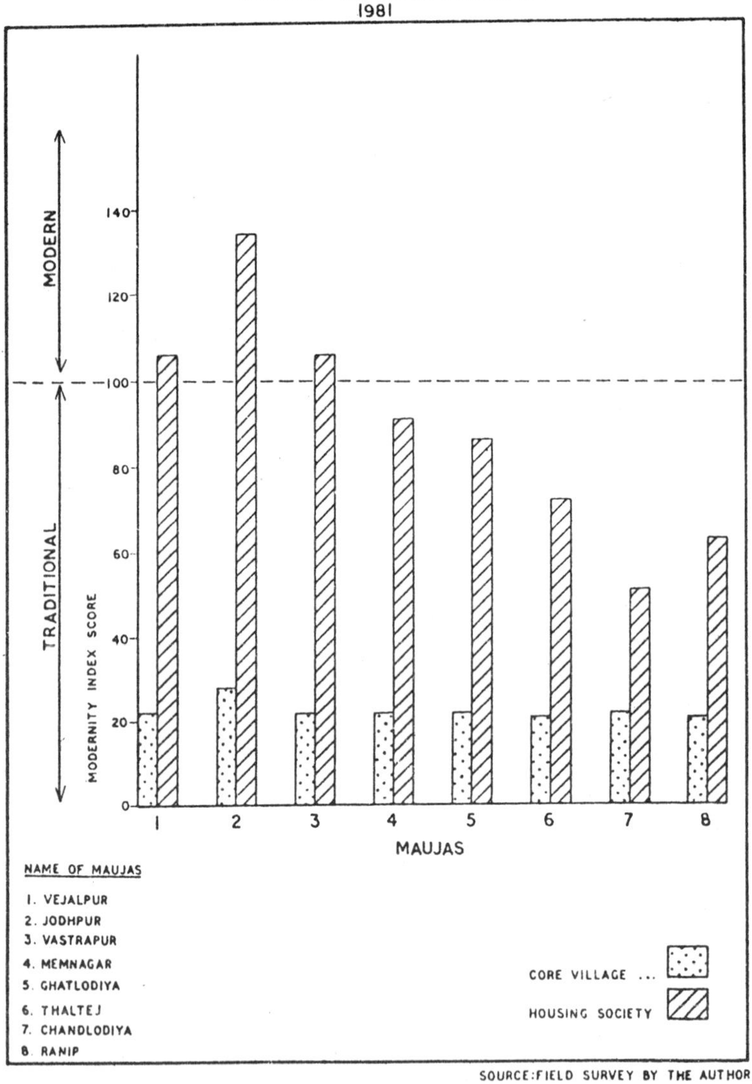

SOURCE: FIELD SURVEY BY THE AUTHOR.

(b) *Reading and recreation habits*

(i) *All core villages versus all housing societies*

Like any other rural population, the residents of core villages do not have reading and recreation habits, which indicate their traditionality. Very few (8%) of them keep newspaper in their houses, and those who keep newspaper, they prefer the newspaper of their own mother tongue (87%). None of the residents of core villages either keep any magazine or the member of any library (Appendix 4F, Table 21-29). They are not member of any club for the purpose of recreation. The younger generation of core villages goes to cinema but older generation does not have any sorts of recreation. In housing societies of most of the maujas, low score in 'Modernity Index' indicates that people of housing societies also have no habit of reading and recreation. 77% of residents of housing societies keep newspaper and very few of them keep newspaper in English (18%) or Hindi (3%) other than their own mother tongue. Buying magazines is also very uncommon habit of the residents of housing societies. The residents generally keep magazines of general information and these are mainly in mother tongue (87%). Some of the residents (57%) read magazines from the circulating library. These circulating libraries are common in the city but not in the suburbs (Figure 8.38).

(ii) *Analysis between core villages*

There is little variation in reading and recreation habits between the residents of core villages of different maujas. Relatively large number of residents (47%) of the core village of Jodhpur keep newspaper in their home. Very few residents of other core villages keep newspaper at home (Appendix 4F, Table 21).

(iii) *Analysis between housing societies*

Major variation between the residents of housing societies regarding their reading and recreation habits is that in the housing societies of Jodhpur, large number of residents (68%) are members of various clubs, i.e. Ellisbrldge Gymkhana, Sports Club, etc., which are associated with high status in social circle. No resident of housing societies of Thaltej, Chandlodiya and Ranip is a member of any club. Few residents of housing societies of Thaltej, Chandlodiya and Ranip keep newspaper, magazine and are member of any library, compared to the residents of housing societies of Jodhpur (Appendix 4F, Table 21-29).

(c) *The behaviour in raising children*

(i) *Analysis between core villages and housing societies*

In some cases there are less differences in score of 'Modernity Index' between the residents of core villages and that of housing societies (Figure 8.39). In can be said that there is little difference between the residents of the core villages and the housing societies in their behaviour of raising children (Appendix 4G, Table 30-36). Traditional style has been indicated by their preference of higher education for their male children than that of the female ones in both the cases (Appendix 4G, Table 30,31). It is also interesting to note that all the residents of core villages and majority (75%) of the people of housing societies give 'caste' the first preference while selecting mate in their children's marriage-another indicator of traditionality (Appendix 4G, Table 32). People with high socio-economic status also give preference to 'caste' as a criterion for selection of mate for their children. There is very little difference bewcen the residents of core villages and housing societies, regarding the number of diseases that a child (below 10 years of age) has experienced (Appendix 4G, Tabel 34). Most (87%/85%) of them had less than four diseases. But there is a difference between the residents of core villages and housing societies regarding the type of treatment being given to children. In housing societies 82% residents prefer allopathic treatment for their children but people in the core villages prefer medical treatment by both trained (53%) and un-trained (32%) doctors. So, some people (23%) in the core villages have not used vaccination of small pox, triple-antigen on their children.

(ii) *Analysis between core villages*

The behaviour of residents of all core villages is not same. The residents of the core villages, except Vejalpur and Jodhpur, expect

Fig. 8.39

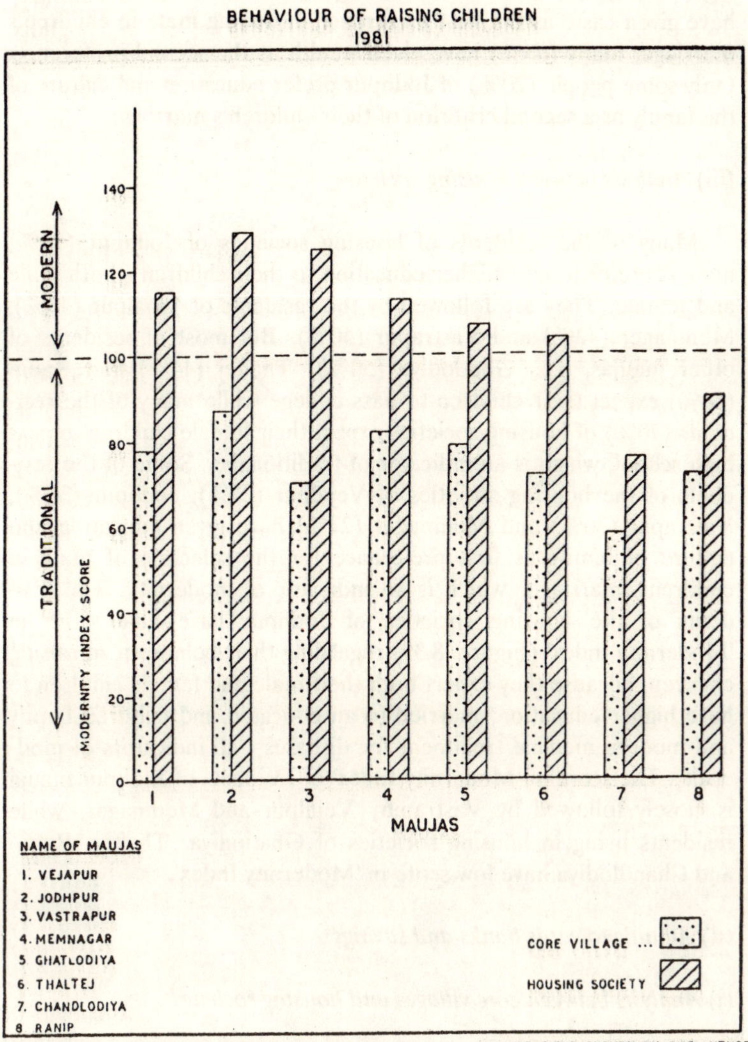

SOURCE: FIELD SURVEY BY THE AUTHOR.

their male child to pass high school and for female child they do not have any fixed expectation. Some (15%) residents of Vejalpur expect graduate degree for both male and female children, while in Jodhpur some (7%) residents expect their children, both male and female, to have higher education. This indicates that the residents of core villages are modern in attitude. Though all residents of core villages have given caste as the first preference in finding mate in children's marriage, many (63%) have given wealth as the second preference. Only some people (20%) of Jodhpur prefer education and culture of the family as a second criterion of their children's marriage.

(iii) *Analysis between housing societies*

Many of the residents of housing societies of Jodhpur (74%) maujas prefer to give higher education to their children - both male and female. They are followed by the residents of Vajalpur (49%), Memnager (42%) and Vastrapur (40%). But most of residents of other maujas, i.e. Ghatlodiya (50%), Thaltej (44%) and Ranip (56%) expect their children to pass college while many of the residents (70%) of housing societies expect their female children to pass high school, which is an indicator of traditionality. Some of the residents of the housing societies of Vejalpur (22%), Jodhpur (26%), Vastrapur (36%) and Memnagar (27%) have given education and culture of family as first preference for the selection of mate in children's marriage, which is an indicator of modernity. The residents of the housing societies of Jodhpur have high score in 'Modernity Index' (Figure 8.39) regarding their behaviour in raising children, because they expect both their male and female children to have higher education, married to an educated and cultured family and modern medical treatment for diseases - all indicators of modernity. The score in 'Modernity Index' of residents of Jodhpur mauja is closely followed by Vastrapur, Vejalpur and Memnagar, while residents living in housing societies of Ghatlodiya, Thaltej, Ranip and Chandlodiya have low score in 'Modernity Index'.

(d) *Attitude towards banks and savings*

(i) *Analysis between core villages and housing societies*

Though investment of money is the natural tendency among the Gujarati community (Appendix 4H, Table 37-40), the people of core villages invest their money in the old traditional way, i.e. buying property and gold, and they have negative attitude towards saving

Fig. 8.40

ATTITUDE TOWARDS BANKS AND SAVINGS
1981

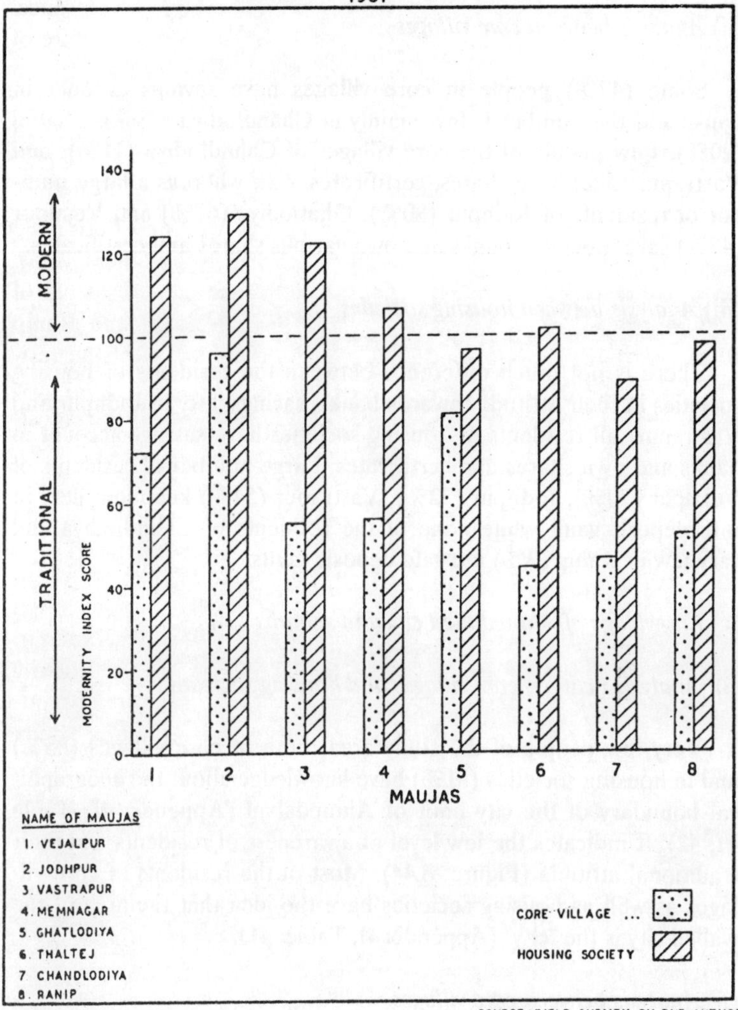

SOURCE: FIELD SURVEY BY THE AUTHOR.

money in the bank or buying shares and certificates. This attitude indicates their traditional behaviour (Figure 8.40). Very few people

(5%) of core villages of all maujas use safe deposit vault in banks to keep jewellery (Appendix 4H, Table 40). The residents of housing societies invest money both in traditional and modern ways. Some (38%) of them keep jewellery in safe deposit vault.

(ii) *Analysis between core villages*

Some (47%) people in core villages have savings account in banks and the number is few mainly in Chandlodiya (28%), Thaltej (20%). Few people of the core villages of Chandlodiya (11%), and Vastrapur (5%) have shares, certificates, etc., whereas a large number of residents of Jodhpur (80%), Ghatlodiya (67%) and Vejalpur (47%) save money in banks and own various shares and certificates.

(iii) *Analysis between housing societies*

There is not much difference between the residents of housing societies in their attitude towards banks, savings, etc. In Jodhpur and Vastrapur, all residents of housing societies have savings account in banks and own shares and certificates. Large number of residents of Vejalpur (70%), Jodhpur (72%), Vastrapur (54%) keep jewellery in safe deposit vault while none of the residents of Chandlodiya and very few in Ranip (3%) use safe deposit vaults.

(c) *Knowledge of Ahmedabad city and suburbs*

(i) *Analysis between core villages and housing societies*

Very few people of the study area, both in core villages (85%) and in housing societies (81%) have knowledge about the geographical boundary of the city limit of Ahmedabad (Appendix 4I, Table 41, 42). It indicates the low level of awareness of residents and their traditional attitude (Figure 8.44). Most of the residents of core villages as well as housing societies have the idea that the area of the walled city is the "city" (Appendix 4I, Table 41).

(ii) *Analysis between core villages*

There is very little differences between the residents of core villages and of suburbs regarding their geographical knowledge of Ahmedabad city and suburbs. Though some people of core villages of Memnagar (58%), Thaltej (47%) do have idea of true city limit, none of the residents of core villages of Vastrapur, Ranip, Chandlodiya have any knowledge about the extension of suburb or what exactly is meant by suburbs (Appendix 4I, Table 42).

Fig. 8.41

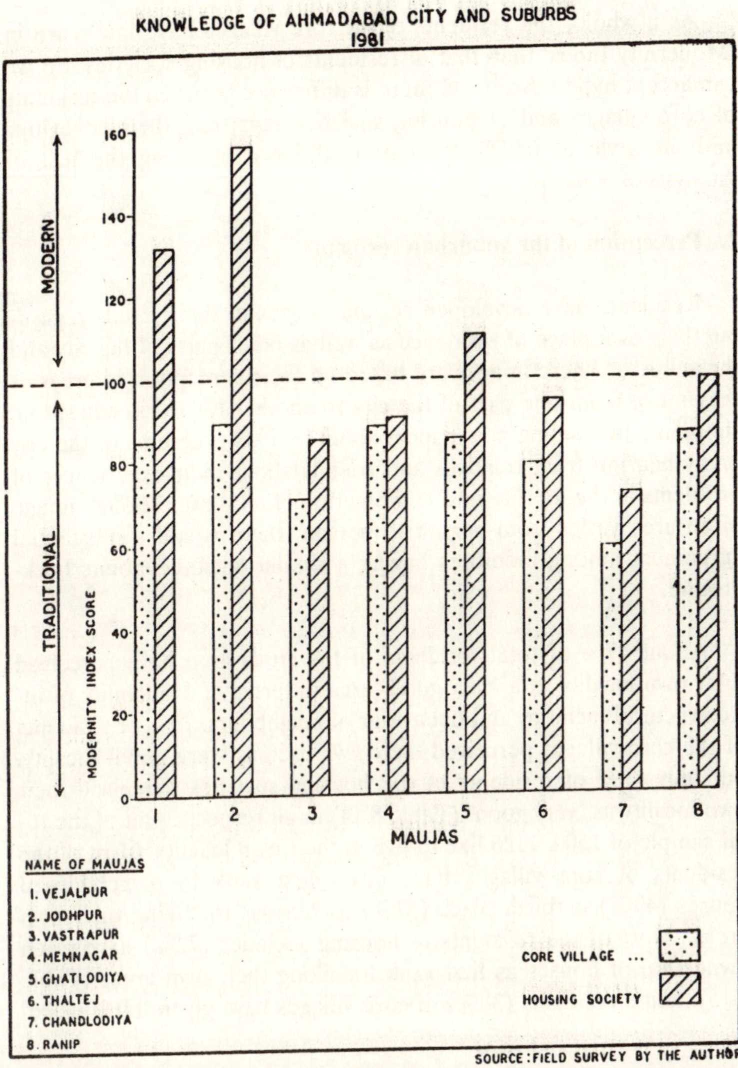

SOURCE: FIELD SURVEY BY THE AUTHOR.

(iii) *Analysis between housing societies*

A large number of residents of housing societies of Jodhpur (47%) and Vastrapur (39%) have idea of the extension of true city and suburban limit. Very few people of housing societies of other maujas have idea of the extension of city and suburbs.

As a whole, the residents of all core villages have low score in 'Modernity Index' than that of residents of housing societies. So we can accept hypothesis 6, i.e. there is difference between the residents of core villages and of housing societies regarding their behaviour and life style at 0.05% level of confidence by using the test of Analysis of Variance.

2. Perception of the suburban residents

Residents have developed certain images in their minds regarding their own place of residence as well as other parts of the city and the suburbs. These images are based on their perception of areas as they move from one part of the city to another for work, education, shopping, by reading newspapers about various problems of the city or by hearing from relatives and friends about their experiences of problems in the locality or a combination of all these. Though image of an area varies from person to person, there is still a substantial agreement among members having a similar socio-economic background.

About 47% of total residents of the study area have perceived their own locality as a "very good" area in terms of availability of infrastructural facilities and behaviour of neighbours. 79% or residents of the core villages perceived their own area as "very good" locality, but only 41% of residents of the housing societies perceived their own locality as "very good" (Table 8.14) in all respects. Out of the total sample of 1500, 1126 like to live in their own locality. Most of the residents of core villages have given first rank to ownership of houses (49%) or birth place (32%) as reasons for liking their own areas. Most of the residents of housing societies (72%) have given ownership of houses as first rank for liking their own areas (Table 8.15). Some residents (38%) of core villages have given "birth place" as second rank of preference, while nearness to friends and relatives is also given the second preference by residents (39%) of core villages. Most of the residents of housing societies have given second rank to good neighbours (50%) and spacious accommodation (28%) as reasons for liking their own areas. This table explains that the residents of core villages like their own area as it is their place of birth, while residents of housing societies, who have migrated here, like their locality due to ownership of house or having good neighbours mainly.

TABLE 8.14: **Resident's perception about their own locality, 1981 (in percentage)**

Type of residence	Nature of Perception	1	2	3	4	5	6
		Very good	Good	Fair	Poor	Very poor	Total %
Core village		79	20	1	-	-	100
Housing societies		41	43	15	1	-	100

Source: Field survey by the author.

Of the total sample, 20% do not like to live in their own area. The main reason (80%) for not liking their own area is lack of various amenities, and (75%) lack of sanitation, garbage removal, etc. The residents of housing societies give more stress on lack of different amenities and facilities for disliking their own locality. Most (80%) of the residents of core villages who dislike their own area have no specific reason (75%). They do not like own area due to the inmigration of the people of high status as neighbours. They are afraid of their cultural identity being engulfed by them. (Table 8.16).

These dissatisfied residents (80%) are not leaving their present locality due to high rent in other areas, not getting suitable house (42%) and not getting house due to non-vegetarian food habit (35%) (mainly for the residents of housing societies) (Figure 8.42).

TABLE 8.15: **Reasons for liking own area of residence (in percentage), 1981**

A - First Rank

	1	2	3	4	5	6	7	8	
Type of residence	Reasons	own house	Birth place	Low Rent	Closer to friends & relatives	Spacious accommodation	Good neighbour	Peaceful area	Total percentage
Core village	49	32	3	16	-	-	-	100	
Housing societies	72	-	18	5	3	1	1	100	
B - Second Rank									
Core village	-	39	-	39	5	17	-	1000	
Housing societies	-	-	-	16	28	50	6	100	

Source: Field survey by the author.

TABLE 8.16: **Reasons for disliking own area of residence, 1981 (in percentage**

A - First Rank

Reasons / Type of residence	1 Lack of various amenities	2 Lack of sanitation, garbage removal etc.	3 Not matching with the status of neighbour	4 No specific reason	5 Total percentage
Core Villages	17	3	-	80	100
Housing Societies	50	13	35	2	100
B - Second Rank					
Core Villages	10	7	75	8	100
Housing Societies	-	67	14	19	100

Perception of the City

It is interesting to note that the residents of core villages and housing societies do not differ very much regarding the choice of residential area. Majority of them prefer their immediate city area as the best place of residence (Figure 8.43). The residents of core villages and housing societies have little difference regarding the perception of various parts of the city. They prefer walled city for shopping and Ashram Road for entertainment, e.g. cinema, restaurant, etc., Navrangpura area for education. They prefer Ashram Road and Ellisbridge area as the best locality of Ahmedabad city (Table 8.17). Suburban people consider the walled city and industrial areas in the eastern parts as congested, dirty and noisy places.

It is interesting to note that the residents of core villages and housing societies do not differ very much regarding the choice of residential area and perception of different parts of the city but they differ considerably regarding their own locality with respect to various aspects like availability of amenities, shopping facilities in the locality, behaviour of neighbours, etc. The residents of the core villages have very high opinion about their own locality. This is mainly due to their more attachment to their places of birth than the residents of housing societes (Table 8.18). The suburban population, due to their attachment to their own areas, have over-rated their own areas. They have high opinion about their area compared to that of the investigator (Table 8.19).

Fig. 8.42

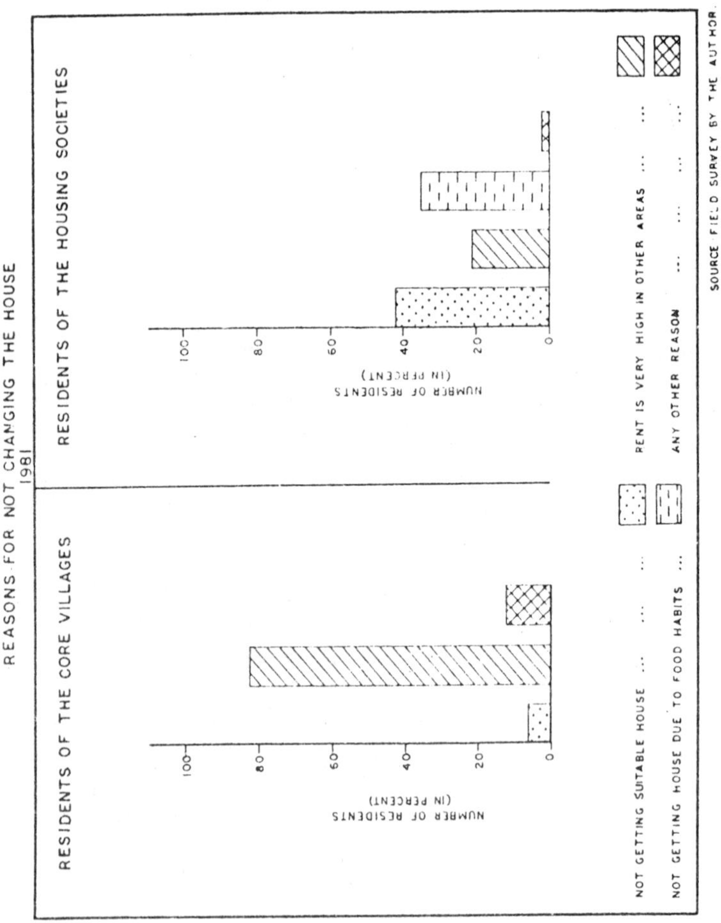
REASONS FOR NOT CHANGING THE HOUSE
1981
RESIDENTS OF THE CORE VILLAGES
RESIDENTS OF THE HOUSING SOCIETIES
NUMBER OF RESIDENTS (IN PERCENT)
0
20
40
60
80
100
NOT GETTING SUITABLE HOUSE
NOT GETTING HOUSE DUE TO FOOD HABITS
RENT IS VERY HIGH IN OTHER AREAS
ANY OTHER REASON
SOURCE : FIELD SURVEY BY THE AUTHOR

TABLE 8.17: The preference of residents to various parts of Ahmedabad city for different activities (in percentage) - 1981

A - Core Villages

Type of activities / Parts of city	1 Walled City Road	2 Ash-ram pura etc.	3 Nav-rang-	4 Nara-pura	5 Paldi Kan-karia	6 Mani-nagar	7 Shahi-bag	8 Mill area	9 Total percen-tage
1. Shopping	99	1	-	-	-	-	-	-	100
2. Entertainment	48	52	-	-	-	-	-	-	100
3. Education	12	.13	75	-	-	-	-	-	100
4. Residential	-	-	32	38	8	6	16	-	100
5. Best area	-	75	17	2	2	-	4	-	100

B - Housing Societies

Type of activities / Parts of city	1 Walled City Road	2 Ash-ram pura etc.	3 Nav-rang-	4 Nara-pura	5 Paldi Kan-karia	6 Mani-nagar	7 Shahi-bag	8 Mill area	9 Total percen-tage
1. Shopping	88	8	2	1	1	-	-	-	100
2. Entertainment	39	59	-	-	-	1	1	-	100
3. Education	10	22	68	-	-	-	-	-	100
4. Residential	2	6	32	28	26	4	2	-	100
5. Best area	-	77	12	7	3	-	1	-	100

Source: Field survey by the author.

Fig. 8.43

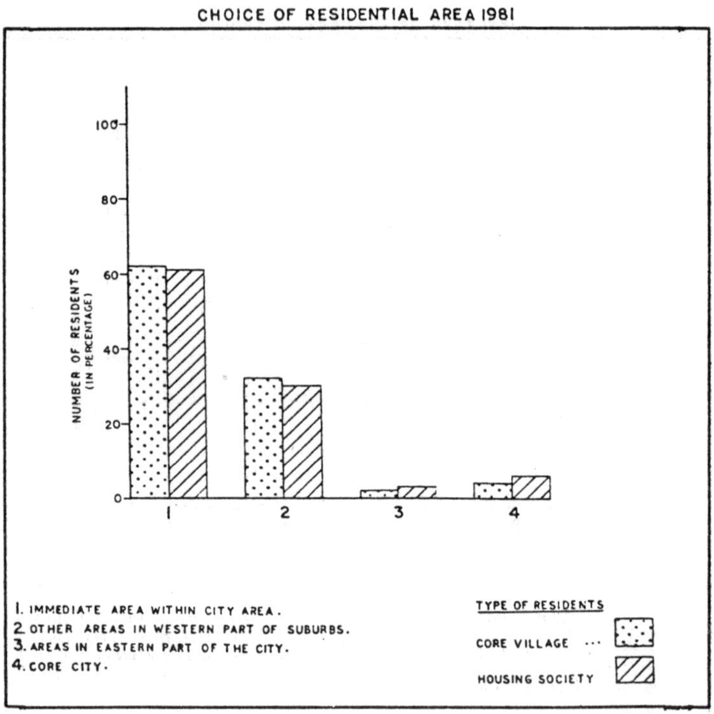

3. **Attachment of suburban residents**

Residents' closeness or tie to their own locality can be judged through a set of variables. A combination of variables like relation with neighbour, considering present locality as neighbourhood, considering own mauja as the best suburb, considering own locality as a social place, duration of stay in the present locality, place of birth in the same locality, nearness of workplace, friends and relatives in own locality, have been combined to form the 'Attachment Index'. The residents of core villages have high attachment to their own locality which indicates their traditional behavriour. So the residents of core villages have higher 'Attachment Index' than the residents of housing societies (Table 8.20). The majority of the residents of the core villages consider their own locality as neighbourhood as they are living there for more than fifteen years. In fact, in many cases, it is for generations that they are living there. Most of them work within their locality as farmers, agricultural labourers, shopowners, etc. and they have their friends and relatives there.

TABLE 8.18: **Evaluation of residents of environmental quality about their own locality, 1981**

Name of maujas / Types of residents	Core villages	Housing societies
1. Vejalpur	0.96	0.87
2. Jodhpur	0.97	0.86
3. Vastrapur	0.91	0.75
4. Memnagar	0.88	0.91
5. Ghatlodiya	0.94	0.68
6. Thaltej	0.98	0.78
7. Chandlodiya	0.88	0.96
8. Ranip	0.85	0.97

Scale of perception index

1) More than 0.95 - Very high

2) 0.91 to 0.95 - High

3) 0.86 to 0.90 - Average

4) 0.81 to 0.90 - Low

5) Less than 0.81 - Very low

Source : Field survey by the author.

Perception index = $\frac{O.P.}{M.P.}$ O.P. = Obtained point

M.P. = Maximum point

Note: Perception index has been measured with respect to the evaluation of residents of their own area regarding available amenities, housing characteristics, and overall nature of the area.

TABLE 8.19: **Evaluation of residents and the investigator about suburban environment, 1981**

Name of maujas / Types of respondents	Residents	Investigator
1. Vejalpur	0.88	0.56
2. Jodhpur	0.83	0.42
3. Vastrapur	0.71	0.38
4. Memnagar	0.67	0.38
5. Ghatlodiya	0.62	0.41
6. Thaltej	0.82	0.25
7. Chandlodiya	0.71	0.28
8. Ranip	0.70	0.33

Scale of perception index

1) More than 0.80 - Very high
2) 0.71 to 0.80 - High
3) 0-61 to 0.70 - Average
4) 0.51 to 0.60 - Low
5) Less than 0.51 - Very low

Source : Field survey by the author.

Perception Index = $\frac{O.P.}{M.P.}$ O.P. = Observed points
M.P. = Maximum points

Note: Perception index has been measured with respect to the evaluation of residents and the investigator of a locality regarding available amenities, housing characteristics and general appearance of the area.

On the other hand, residents of housing societies are less attached to their own locality. They have recently migrated to the suburbs and have not spent enough time to have emotional feelings for the place. They go to the different parts of the city for work, to meet friends and relatives and many of them do not consider their present locality as their own neighbourhood. So, obviously, the residents of housing societies have less score in 'Attachment Index'. The residents' attachment to their own locality has a positive relationship with the perception of their own locality (Table 8.14). The residents of core villages due to their high attachment to their own locality have over-rated the environmental quality of their own area than the residents of housing societies.

Fig. 8.44

INTERACTION OF RESIDENTS
1981

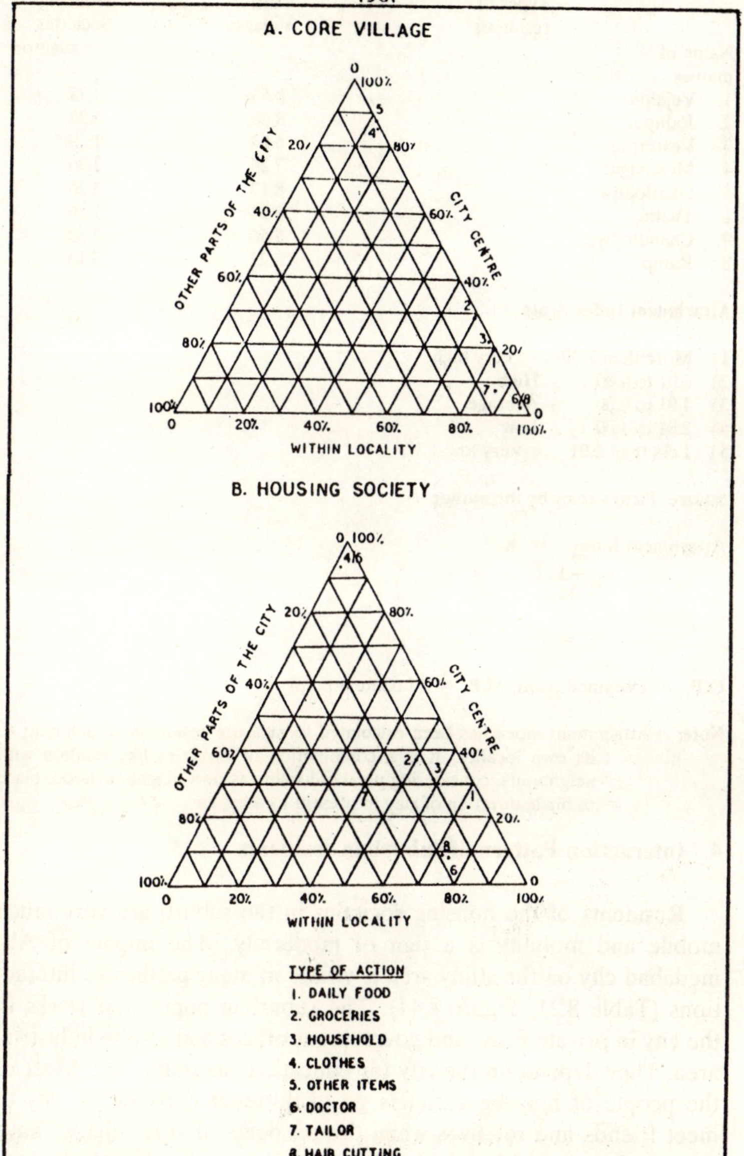

SOURCE: FIELD SURVEY BY THE AUTHOR.

TABLE 8.20: **Attachment of residents, 1981**

Name of maujas / Types of residents	Core Villages	Housing Societies
1. Vejalpur	8.53	3.75
2. Jodhpur	8.60	3.25
3. Vastrapur	6.84	1.29
4. Memnagar	7.26	2.36
5. Ghatlodiya	8.17	3.26
6. Thaltej	8.24	1.26
7. Chandlodiya	5.56	2.52
8. Ranip	6.22	3.13

Attachment Index Scale

1) More than 8.00 - Very high
2) 6.01 to 8.00 - High
3) 4.01 to 6.00 - Average
4) 2.01 to 4.00 - Low
5) Less than 2.01 - Very low

Source: Field survey by the author

$$\text{Attachment Index} = \sum_{i=1}^{n} \frac{\Sigma \text{O.P.}}{\text{M.P.}}$$

O.P. = Obtained point; M.P. = Maximum point

Note: Attachment index has been calculated to measure residents' attachment to its own locality. It is a combination of variables like relation with neighbours, considering present locality as own neighbourhood, place of birth, duration of stay in present locality, etc.

4. Interaction Pattern of suburban residents

Residents of the housing societies in the suburb are very much mobile and mobility is a sign of modernity. The impact of Ahmedabad city on the study area is visible in many pattern of interactions (Table 8.21, Figure 8.44). The suburban population works in the city in private firms and government offices and also in industrial area. They depend on the city for education, shopping, etc. Most of the people of housing societies go to different parts of the city to meet friends and relatives while the residents of core villages have more interaction within the locality regarding work, education, meeting of friends and relatives. So, the residents of core villages are spatially restricted, which is a sign of traditionality. Regarding recreation, most of the suburban population go to cinema while the residents of core villages go to religious places. This is another sign of traditionality. More than 50% of the suburban population go to many parts of the city, mainly to the city-centre for shopping clothes and for buying sophisticated and expensive things. But they prefer to have their doctors, saloon, tailors within the locality (Figure 8.44). As a whole, residents of core villages depend more on local market for shopping than the residents of housing societies.

TABLE 8.21: **Interaction pattern of residents (in percentage) - 1981**

A - Core Villages

Type of actions / Various areas	1 Within locality	2 City centre of the city	3 Other parts	4 Total percentage
1. Work	89	3	8	100
2. Education	83	1	16	100
3. To visit friends	89	1	10	100
4, To visit relatives	92	-	8	100
5. Recreation	20	43	37	100

B - Housing societies

Type of actions / Various areas	1 Within locality	2 City centre of the city	3 Other parts	4 Total percentage
1. Work	1	47	52	100
2. Education	6	15	79	100
3. To visit friends	18	16	66	100
4. To visit relatives	1	41	58	100
5. Recreation	1	28	71	100

Source: Field survey by the author.

5. Summary of the chapter

Regarding the attitude, behaviour and life style, residents of core villages of all maujas are traditional, i.e. they do not have any reading habit, they do not keep any magazine or newspapers; they have very little desire about children's education. They always prefer caste as a first criterion in children's marriage. The residents of core villages possess and use very few modern goods while residents of housing societies have relatively modern behaviour and life style. They expect higher education for their children, they possess and use various modern goods but most of them give caste as a first preference in the marriage of their children. As a whole, behaviour and life style of the residents of housing societies is not as modern as their socio-economic status. Their level of education, occupation and income are high, but their attitude towards life has not changed so much.

There is also difference in perception, attachment and interaction pattern of the residents of core villages and housing societies. Residents of core villages have rated their own area very high as they have high attachment to their own area. As a whole, residents of the suburbs have rated their area higher than that perceived by the investigator. Due to the emotional feeling of their own locality for generations, the residents of core villages are more attached to their own area. Regarding interaction patterns, residents of housing societies depend more on the city for shopping, social visits, work, education and recreation than the residents of core villages.

NINE

THE HOUSING CHARACTERISTICS OF THE SUBURBAN RESIDENTS

The housing characteristics of the suburban population have analysed with the help of nine variables (Appendix 4 J-M. Table 43-51). These nine variables have been grouped into four sub-factors, i.e. housing structure, housing facilities, value of the houses and internal density. All data used in this analysis have been collected from the field survey. In this Chapter, analysis will be done, comparing housing characteristics, first, between all core villages and all housing societies and then between core villages and housing societies of different maujas. These comparisons will be done with the help of 'Modernity Index' as it has been done in the case of 'Socio-economic status' and 'behaviour and life style' of residents. The houses made of cement, concrete and burnt brick; with facilities of separate bathroom, toilet, kitchen and water tap; high rent and high selling value; and low internal density have been termed "modern".

Core villages, as a whole, are original residential areas of study area. So, they have traditional type of houses. The score in 'Modernity Index' of all sub-factors are low in all core villages. But their score is very low in case of housing facilities and value of the house, because most of the houses in the core villages do not have any facilities like separate bathroom, toilet, water tap, and the value of the houses is also low. So, the rent of the houses in the core village is lower than that of housing societies and their selling value is also low. As in other factors, there are differences within core villages and within housing societies situated in various maujas.

1. Housing Structure

(a) *Core villages versus all housing societies*

The variables like the materials used for roof and wall, number of rooms (without kitchen), and built up area have been combined to form the housing structure (Appendix 4J, Table 43-46). In core villages, there are some houses (28%) which are made of grass, leaves, stones, corrugated iron sheets, etc. The majority (45%) of houses in core villages have brick roof. The smaller number of concrete roofs (17%) in the core villages is mainly due to the older age of the houses in core villages and presence of fragile material like grass, leaves etc. is due to the lower socio-economic status of population. Most (74%) of the houses in core villages have less than three rooms and their size is also less than 80 square yards (89%). But in housing societies all houses have walls of burnt bricks and have concrete roofs - a sign of modernity. In housing societies, the houses are large. So, houses with more rooms are seen frequently (44%) though most (77%) of the houses are less than 80 square yards in built up area.

(b) *Analysis between core villages*

There are much variations in the housing structures between core villages situated in different maujas. In core villages of Chandlodiya and Ranip, no house has concrete roof, while in the core village of Jodhpur a large (40%) number of houses has concrete roofs. In core village of Jodhpur (13%), Vastrapur (10%) and Memnagar (16%) some houses have more than four rooms. In Chandlodiya all houses have less than three rooms. It indicates that number of rooms in a house has some relation with the economic status of the family. Regarding the built up area of houses in core villages, only Ghatlodiya has few (2%) houses with more than 149 square yards of built-up area, but all the houses in core villages of Ranip and Chandlodiya have less than 80 square yards or area.

(c) *Analysis between housing societies*

In the housing societies of Jodhpur (11%) and Memnagar (10%)

some houses have more than four rooms. Only in the housing societies of Jodhpur(11%) have an area of more than 149 sq. yards of built-up area, and majority of houses of Chandlodiya (98%) and Ranip (83%) have less than 80 square yards of built-up area.

There is some relation between the socio-economic status of residents and their housing structure. So, the residents of housing societies have high score in the 'Modernity Index' than the residents of core villages regarding their housing structure (Figure 9.45).

2. Housing facilities

(a) *All core villages varsus all housing societies*

Due to traditional behaviour and life style, most of the houses in core villages are devoid of basic amenities like separate kitchen (68%), bathroom (52%) and toilet (64%). These are indicators of traditionality (Appendix 4K, Table 47). Some of the Houses in core villages do not have electricity (16%) and water taps. In Housing societies, all hosuses have separate electricity and water facilities. Most of the households have separate toilet (98%), bathroom (98%) and Kitchen (99%) in their own houses.

(b) *Analysis between core villages*

All houses in the core villages of Jodhpur and Vejalpur have electricity and most of them have separate water taps, bathrooms and toilet and most (93%) of the houses in core villages of Jodhpur have separate kitchen also (a sign of modernity). On the other hand, in core villages of Ranip, none of the houses has separate kitchen and toilet and separate water tap and bathroom. In core villages of Vastrapur, Thaltej and Chandlodiya also few houses have separate bathroom and toilet.

(c) *Analysis between housing societies*

The only variation between the residents of housing societies of different maujas regarding their housing facilities is that in Vejalpur, Ghatlodiya and Chandlodiya few households share Kitchen, toilet, bathroom within themselves. Sharing is done mainly in the houses where owner and tenant live in the same premises.

Fig. 9.45

HOUSING STRUCTURE
1981

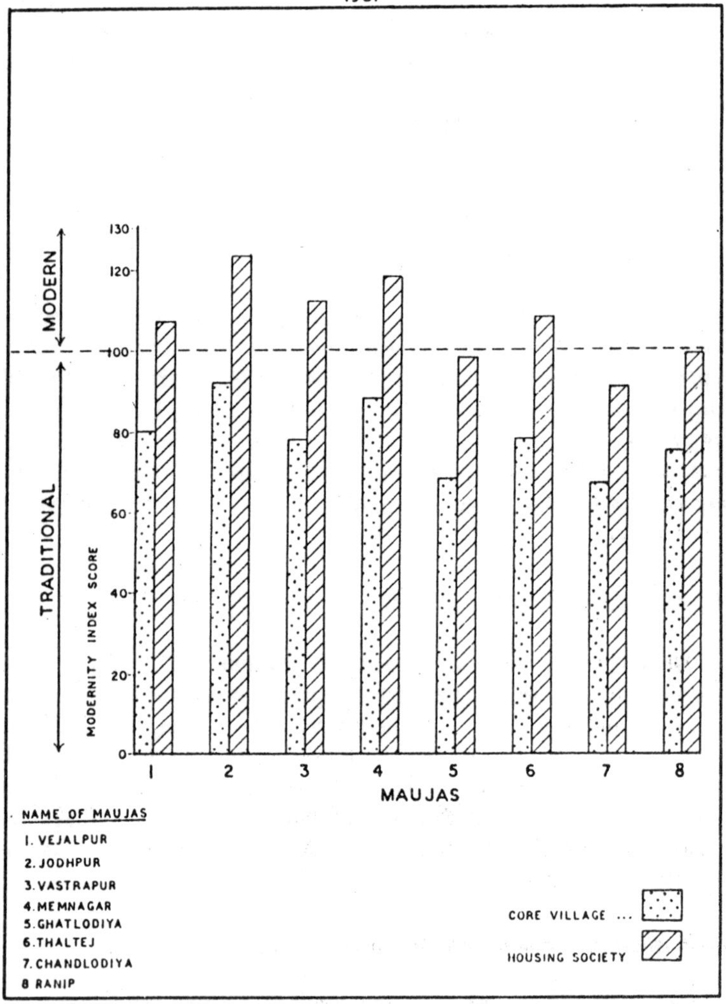

SOURCE : FIELD SURVEY BY THE AUTHOR

Housing facilities of the study area reflects the behaviour and life style of population. The people of core villages, due to their traditional behaviour, do not feel the necessity of separate bathroom, toilet and kitchen in their houses. So, they have low score in the 'Modernity Index' than the residents of housing societies regarding the housing facilities (Figure 9.46). But the unhygenic way of livelihood (i.e. using open space for toilet) of the people of the core villages make the environment of the surrounding area of villages polluted.

3. Value of Houses

(a) *All core villages versus all housing societies*

It is obvious that houses in core villages have low rent and selling value (Appendix 4L, Table 48, 49) because they have low score in 'Modernity Index' both in housing facilities and housing structure. Most of the houses (60%) in core villages have rent less than 100 rupees per month and have selling value less than Rs.60,000 (51%). The houses in the housing societies have higher rent and higher selling value than the houses in core villages. So, they have high score in 'Modernity Index' than the houses in core villages (Figure 9.47). Most (70%) of the houses in housing societies have rent more than Rs.199 per month and some of them (17%) have rent more than Rs.399 per month. Though most (56%) of the houses have selling value around Rs.1,00,000 but some (16%) of them have selling value more than Rs.1,49,000.

(b) *Analysis between core villages*

The rent of houses in the core villages have close relationship with the socio-economic aspects of residents. For example, in core villages of Vastrapur and Ranip, where a large number of in-migrated residents are engaged in non-primary activities, have higher rent of houses due to the more demand for rented houses. In Vastrapur, 75% of houses have rent between Rs.100-199 per month. This has been followed by Ghatlodiya (67%) and Chandlodiya (50%). The selling value of houses, which has been assessed by house owners, is high in core villages with higher socio-economic status, i.e. Jodhpur. In Jodhpur, houses are bigger in size, so obviously the selling value is also high, i.e. most (60%) of the houses in core village of Jodhpur mauja have selling value between Rs.60,000-

Fig. 9.46

HOUSING FACILITIES
1981

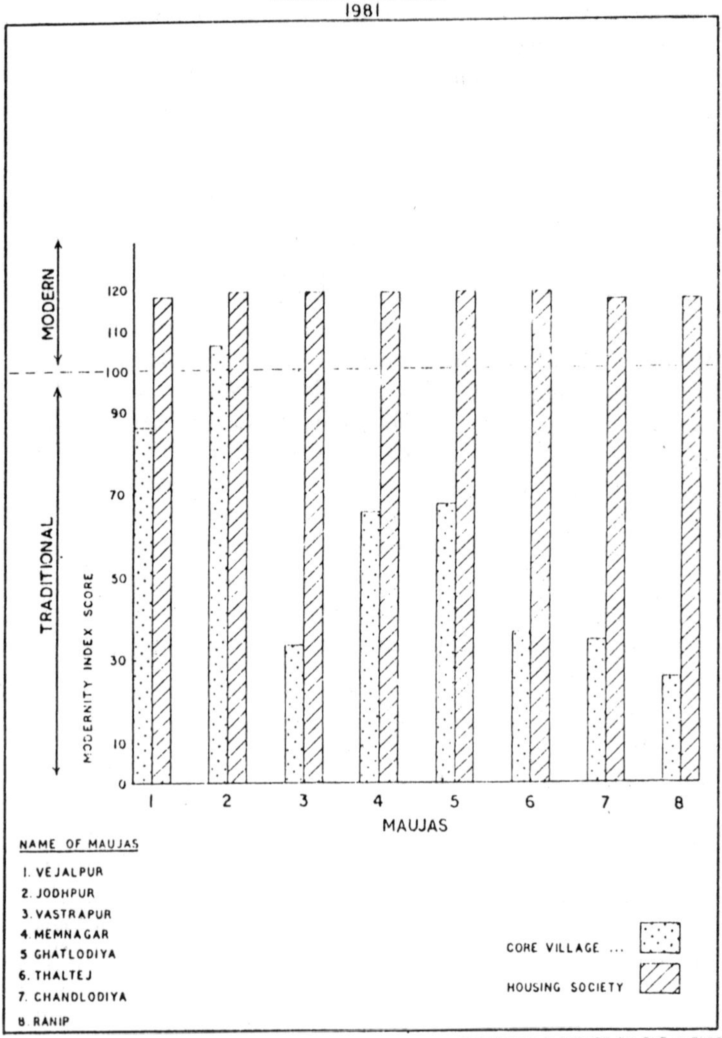

SOURCE: FIELD SURVEY BY THE AUTHOR.

99,000. There are some (7%) houses in the core villages of Jodhpur whose selling value is more than Rs.1,49,000. On the other hand, in core villages of Vastrapur and Chandlodiya, most of the houses have

selling value less than Rs.60,000. This is mainly due to the poor housing structure and poor housing facilities.

Fig. 9.47

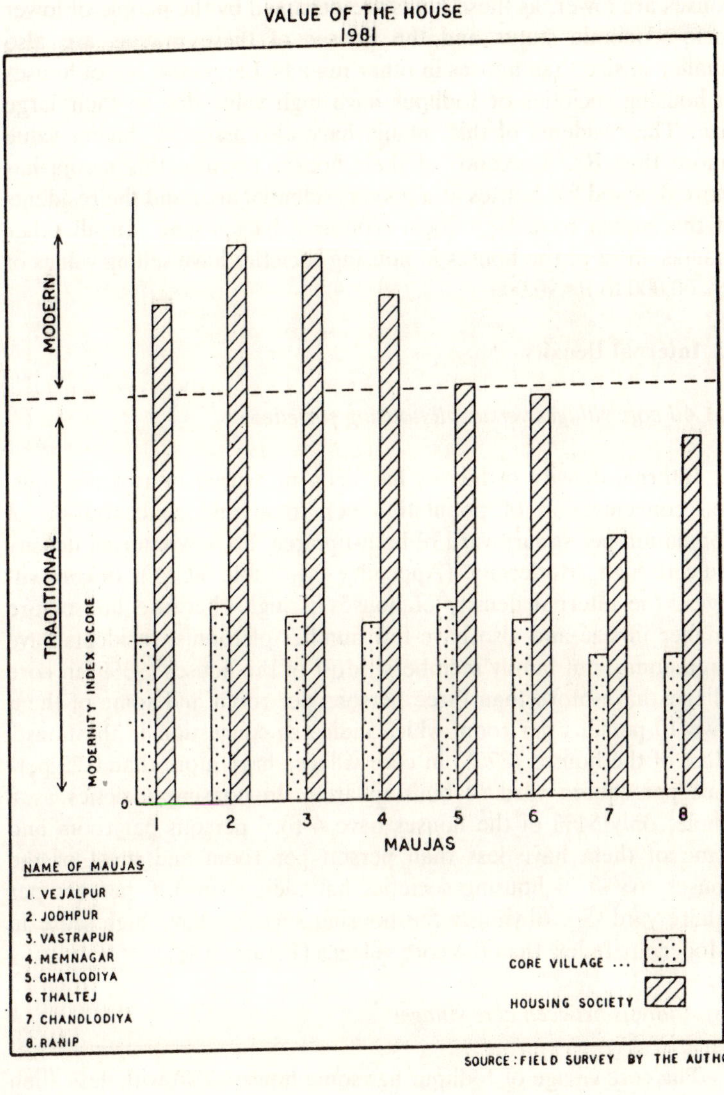

SOURCE: FIELD SURVEY BY THE AUTHOR.

(c) *Analysis between housing societies*

The rent of houses (by room) in housing societies of Vejalpur, Jodhpur, Vastrapur, Memnagar and Ghatlodiya is relatively high (Appendix 4L, Table 40). In Chandlodiya and Ranip, the rent of the houses are lower, as these areas are occupied by the people of lower socio-economic status and the houses of these maujas are also smaller in size than houses in other maujas. Large number of houses in housing societies of Jodhpur have high value due to their large size. The residents of this mauja have also assessed higher value (more than Rs. 1,oo,ooo) of their houses because this mauja has more demand for houses at a good residential area and the residents of this mauja have high socio-economic background. In all other maujas, most of the houses in housing societies have selling values of Rs. 60,000 to Rs.90,000.

4. Internal Density

(a) *All core villages versus all housing societies*

Internal density of houses has been measured by two methods, i.e. concentration of population per room and concentration of population per square yard of built-up area. So, low internal density indicate high "Modernity" (Appendix 4M, Table 50, 51). In core villages, the internal density of houses is high because houses are smaller in size and also have few number of rooms, residents have large number of family members. Most of the houses (93%) in core villages have more than three persons per room and some of them have 10 persons per room which indicates congestion in the house. Most of the houses (87%) in core villages have more than 0.29 persons per square yard of built-up area. In housing societies as a whole, only 54% of the houses have 4 to 6 persons per room and some of them have less than person per room and most of the houses (68%) in housing societies have less than 0.10 persons per square yard. So, obviously the housing societies have high score in 'Modernity Index' than the core villages (Figure 9.48).

(b) *Analysis between core villages*

The core village of Jodhpur has some houses(7%) with less then

Fig. 9.48

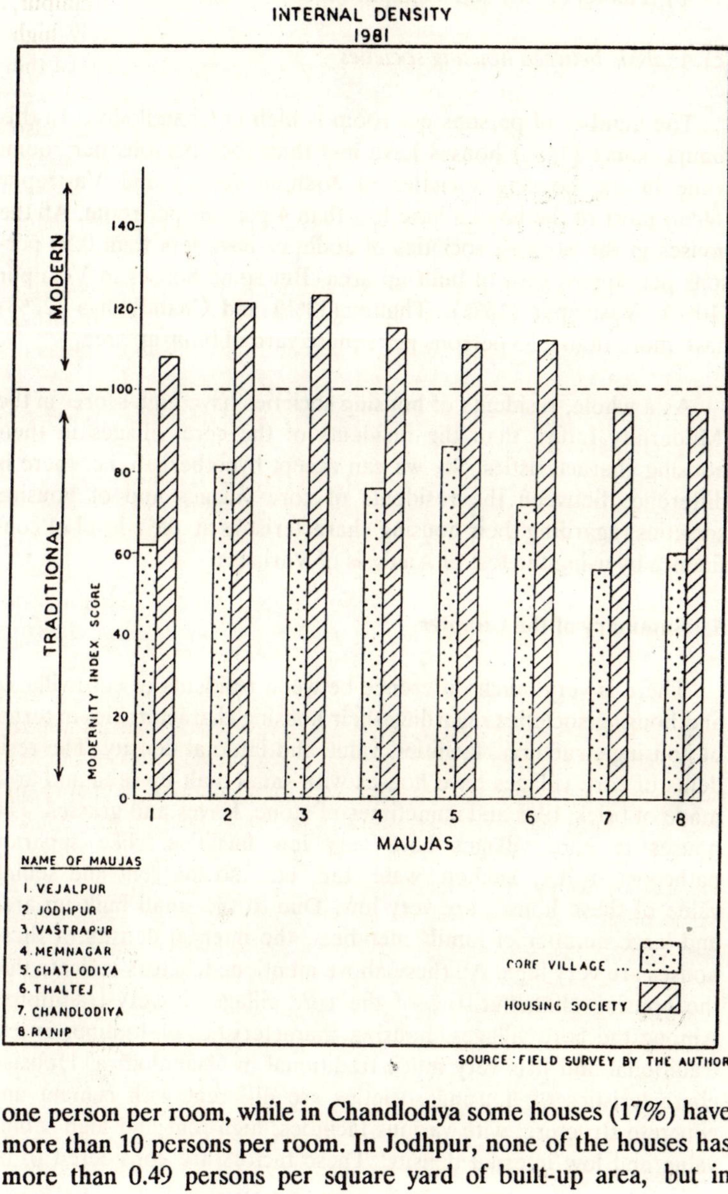

one person per room, while in Chandlodiya some houses (17%) have more than 10 persons per room. In Jodhpur, none of the houses has more than 0.49 persons per square yard of built-up area, but in

Chandlodiya 22% of the households have more than 0.49 persons per square yard. Chandlodiya has been followed by Vastrapur (18%), Thaltej (13%) and Vejalpur (10%).

(c) *Analysis between housing societies*

The number of persons per room is high in Chandlodiya. In this mauja, some (14%) houses have less than four persons per room, while in the housing societies of Jodhpur (66%) and Vastrapur (69%) most of the houses have less than 4 persons per room. All the houses in the housing societies of Jodhpur have less than 0.50 persons per square yard of built-up area. But some houses in Vejalpur (10%), Vastrapur (16%), Thaltej (13%) and Chandlodiya (22%) have more than 0.50 persons per square yard of built-up area.

As a whole, residents of housing societies have high score in the 'Modernity Index' than the residents of the core villages in their housing characteristics. So, we can accept hypothesis 6. i.e. there is difference between the residents of core villages and of housing societies regarding their housing characteristics at .05% level of confidence by using the test of Analysis of Variance.

5. Summary of the Chapter

There is very much difference between residents of core villages and housing societies regarding their housing characteristics in terms of housing structures, facilities, value and internal density. The residents of core villages have houses with small built-up area and area made of brick, tiles and sometimes of stone, leaves and grasses. The houses in core villages have very few facilities, like separate bathroom, toilet, kitchen, water tap, etc., so the rent and selling value of these houses are very low. Due to the small built-up area and large number of family members, the internal density of these houses are very high. All these above mentioned factors indicate that the housing characteristics of the core villages is very traditional. Among the core villages, housing characteristics of Jodhpur is less traditional and it is very much traditional in Chandlodiya. Housing characteristics of housing societies are different with cement and concrete structure, with various facilities, high rent and high selling value and low internal density. These factors are considered to be modern. Housing societies of Jodhpur are very modern which is closely followed by Vastrapur, whereas the level of modernity of the housing societies of Chandlodiya is very low.

TEN

THE SOCIO-DEMOGRAPHIC MODEL

In the previous three Chapters, i.e. VII, VIII and IX, the socio-economic status, behaviour and life style and housing characteristics of suburban population have been analysed to understand the difference between the residents of core villages and housing societies of different maujas. It has been found that all the core villages have traditional socio-economic status, behaviour and life style and housing characteristics but housing societies in most of the maujas have modern socio-economic status and housing characteristics but traditional behaviour and life style of residents. It is interesting to note that people of all core villages are not traditional in the same manner and also the people of housing societies and not modern in the same rate. So, to find out a residential pattern and a typology on space, the 'Socio-Demographic Model' has been formulated using the above mentioned three major factors. The model is represented by a three-dimensional rectangular parallelopiped graph. Three axes of this graph have been represented by socio-economic status (X-axis), behaviour and life style of residents (Y-axis) and their housing characteristics (Z-axis). These three factors are very much interrelated. As for example, socio-economic status which has variables like level of occupation, education, monthly income, family background etc. controls the behaviour and life style, like possession and use of various goods, reading and recreation habit, the style in raising children, their attitude and belief, etc. Both socio-economic status and behaviour and life style control housing characteristics i.e. housing structure, size, facilities, value of the house and internal density, etc. This model measures the level of 'Modernity' of the suburban residents in their socio-economic status, behaviour and life style and housing characteristics. The term 'Modern' used in this study means freedom of thought, an enlightened framework of mind and technological sophistication, whereas the term 'Traditional' means the belief in age-old traditional institutions, a state of mind less conducive to change, and resistance to accept scientific way of life. In this model, all residential structures, i.e. of core villages, of housing societies and of other types of houses, have been combined in a mauja. The mauja has been considered as a single unit to find out the difference between various maujas regarding socio-economic status, behaviour and life style and housing characteristics so that a suburban typology of household and housing can be evolved.

In the following sections, spatial distribution of three factors have been described briefly.

Factor-I : Socio-economic Status

The study area displays a mixed pattern of socio-economic type of community (Figure 10.49). The southern portion of the study area has modern socio-economic status and northern part has traditional status, while the central part is a transitional area having both modern and traditional status giving rise to a mixed pattern.

Vejalpur and Jodhpur situated in the southern part of the study area are the extension of the developed areas of the city. Vejalpur, situated in the extreme south, is one of the earliest developed mauja of the study area. It is also the most developed mauja regarding accessibility, connectivity and amenities. This is the only mauja where a large number of Muslim population of higher socio-economic background had migrated from the core city in housing societies of this mauja.That has made the mauja cosmopolitan in nature. Though the socio-economic status of the residents of the core village is quite traditional, the modern socio-economic status of a large population of housing societies has made the overall score high in 'Modernity Index' in this mauja as a whole.

Jodhpur is the most cosmopolitan mauja of the study area. The original residents of the core village, i.e. Patels of this mauja, have high socio-economic status than the residents of other core villages

That attracted the people of modern socio-economic status to settle in housing societies of this mauja.

Fig. 10.49
FACTOR I: SOCIO-ECONOMIC STATUS

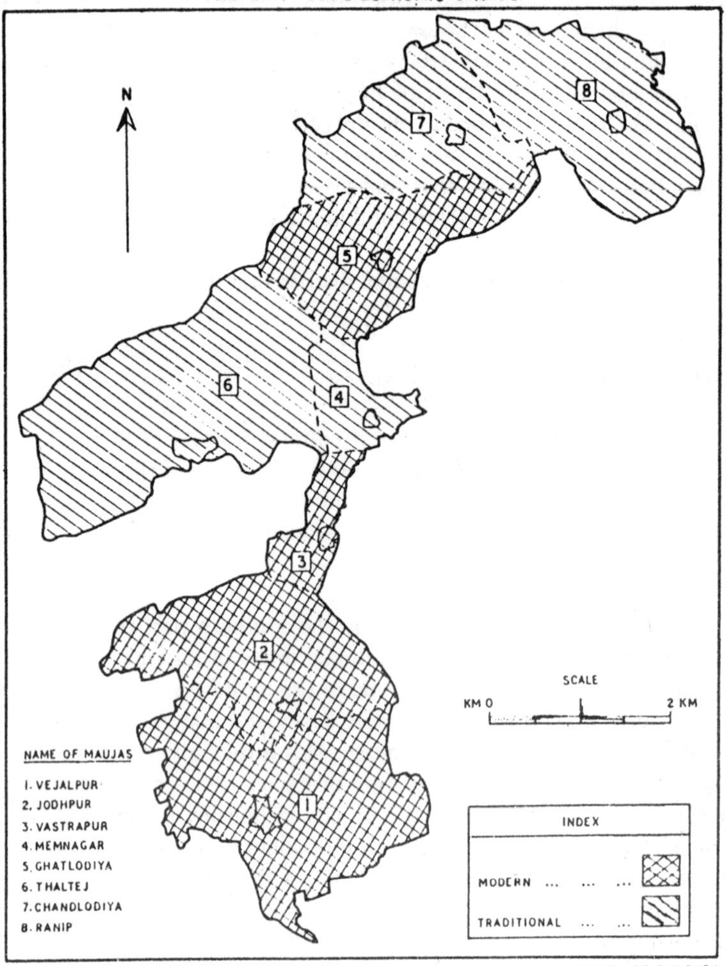

SOURCE: FIELD SURVEY.

Vastrapur has a number of institutions. So, a large portion of its residents, both of core village and housing societies, has modern occupational status as they work in these institutions. Higher level of occupation of residents and presence of large number of inmigrants have increased the socio-economic status of this mauja.

Ghatlodiya, situated in the north-west, also has modern socio-economic status, but the score in 'Modernity Index' is lower than that of Jodhpur. Though till the later half of 1970's Ghatlodiya did not have good amenities and facilities, this mauja has grown very fast recently due to the improvement in accessibility and infrastructural facilities. So, this mauja has been developed as a cosmopolitan area with residents of relatively high socio-economic status in its housing societies.

Ranip and Chandlodiya situated in the extreme north and north-west and Thaltej in the extreme west, have residents with traditional socio-economic status. In Ranip, the industrial landuse, along with poor accessibility and long distance from the city-centre, has made this mauja less attractive to the people of higher socio-economic status. In Chandlodiya and Thaltej, poor connectivity and distance from the city-centre have discouraged the concentration of the people of higher socio-economic status in these maujas. But the concentration of population of traditional residents in Memnagar is an unexpected phenomenon, as Memnagar is the extension of a posh locality of the city. The socio-economic status is the combination of many variables, so it can be said that residents of Memnagar have high score in some variables like monthly income, level of occupation etc. but they have low score in other variables, like type of family, food habit etc. So they have relatively low score in 'Modernity Index' in general in socio-economic status.

So the concentration of residents with high socio-economic status mainly depends on factors like its nearness to the city-centre, high connectivity, availability of amenities and the caste of original residents of core villages.

Facto-II : Behaviour and life Style

There is a contrast in the spatial distribution of the socio-economic status and behaviour and life style of residents. The modern behaviour and life style which has been indicated by the residents' possession and use of modern goods, expectation of higher education for children, reading and recreation habits etc. have been

Fig. 10.50

FACTOR II: BEHAVIOUR AND LIFE STYLE

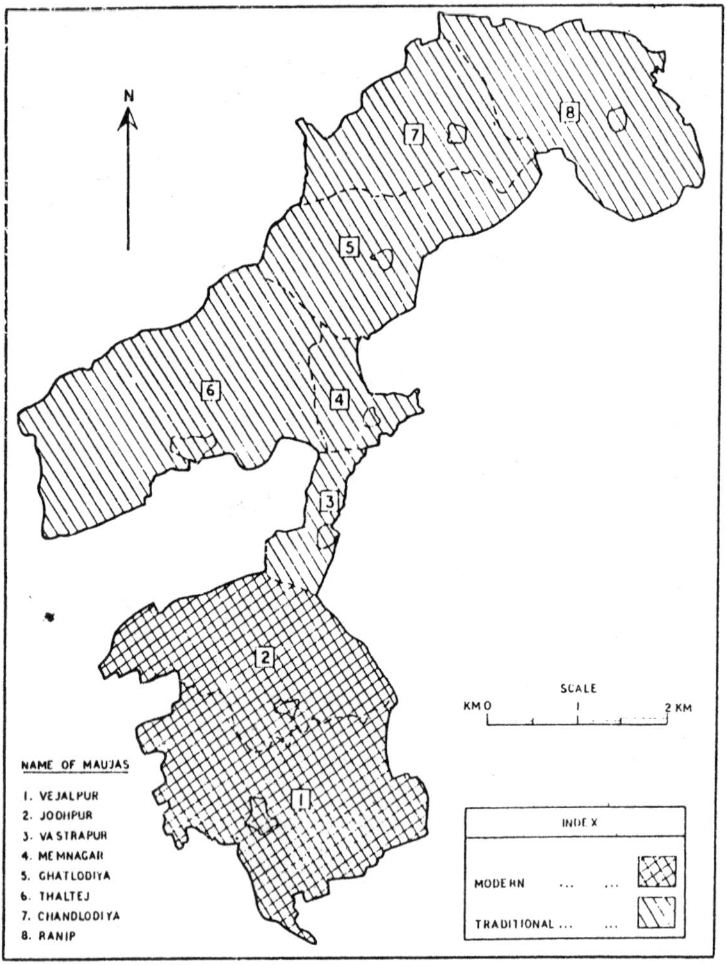

SOURCE: FIELD SURVEY.

present only in two maujas in the extreme south, i.e. Jodhpur and Vejalpur. These two maujas in the extreme south, also have residents with modern socio-economic status. But, in other two maujas, i.e. Vastrapur and Ghatlodiya, where residents are having modern socio-economic status have traditional behaviour and life style. This contrast in spatial distribution is due to the fact that there has been a change in suburbanisation but behaviour and life style of residents have not changed in a same manner.

In four other maujas, i.e. Ranip, Chandlodiya, Thaltej and Memnagar, residents have traditional behaviour and life style and similarly they have low socio-economic status of residents. So there has not been any contrast between two factors (Figure 10.50).

Fig. 10.51

FACTOR III: HOUSING CHARACTERISTICS

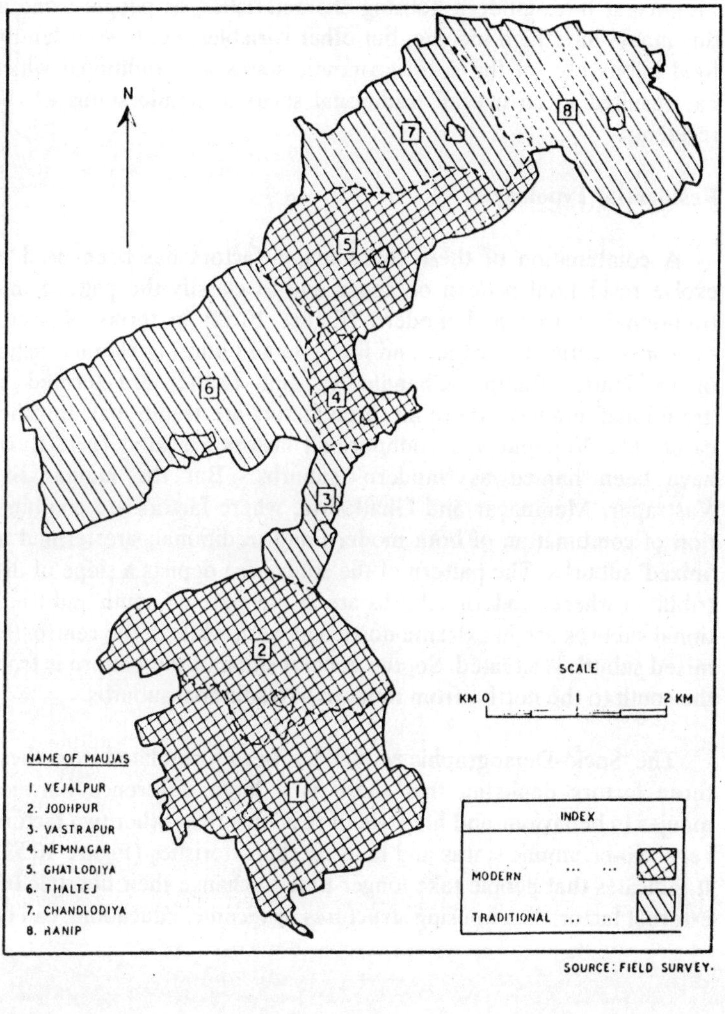

SOURCE: FIELD SURVEY.

Factor-III : Housing Characteristics

The housing characteristics more or less follow the same pattern as socio-economic status except in Vastrapur and Memnagar (Figure 10.51). In Vastrapur, most of the people live in the core village, so, they have traditional housing characteristics, i.e. without various facilities (like separate bathroom, toilet, kitchen), small built-up area, low selling value of houses and high internal density. But as most of them are engaged in higher level of occupation, they have modern socio-economic status. On the other hand, residents of Memnagar have modern housing characteristics, as people living in this mauja have high income, but other variables, i.e. type of family, food habits etc. of the socio-economic status are traditional which has resulted in an overall traditional socio-economic status of the residents.

Residential Typology

A combination of these three major factors has been used to evolve residential pattern on space and to classify the pattern into traditional, mixed and modern (Figure 10.52) in terms of socio-economic status, behaviour and life style and housing characteristics of residents. Ranip, Chandlodiya and Thaltej are termed as 'traditional' maujas, where all three factors are traditional. In other cases, like Vejalpur and Jodhpur, all factors are modern as those have been named as 'modern' suburbs. But the maujas like Vastrapur, Memnagar and Ghatlodiya, where factors are combination of combination of both modern and traditional, are termed as 'mixed' suburbs. The pattern of the study area depicts a slope of distribution where modern suburbs are in the extreme south and traditional suburbs are in extreme north and west, while in the centre, the mixed suburb is situated. So, the general slope of the pattern is from the south to the north - from modern to traditional suburbs.

The 'Socio-Demographic Model' has been formulated from these three factors depicting that there is a large difference between maujas in behaviour and life style of residents than other two factors, i.e. socio-economic status and housing characteristics (Figure 10.53). It indicates that people take longer time to change their life style but external factors like housing structures or income, education, can be

Fig 10.52

SOCIO-DEMOGRAPHIC MODEL

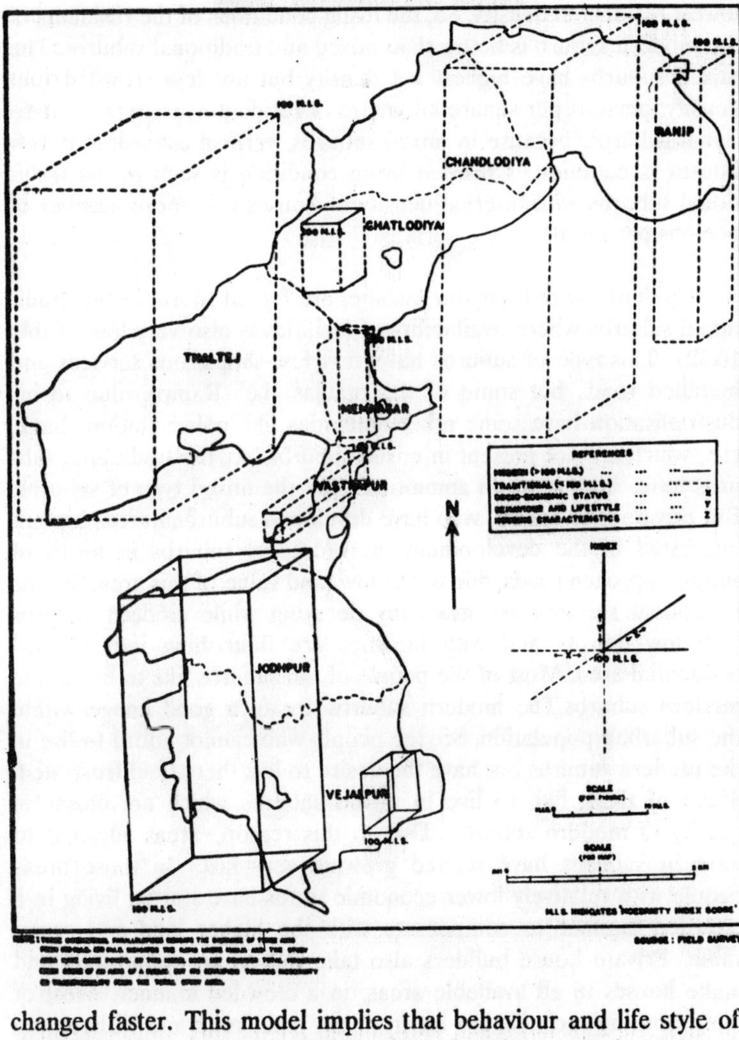

SOURCE : FIELD SURVEY.

changed faster. This model implies that behaviour and life style of people do not change as fast as that of socio-economic status or housing characteristics. This model also helps us to analyse various processes that are responsible for the growth of the suburbs. Modern suburbs (Vejalpur and Jodhpur) have the highest percentage of agricultural and residential area (Table 10.22). The modern suburbs also have the highest growth rate of population (Table 10.22). But it is interesting to note that the modern suburbs have lowest residential density. So, the living conditions of the residents of the modern suburb is better than mixed and traditional suburbs. The mixed suburbs have highest net density but are less crowded (net density: persons per square kilometre of residential area) than traditional suburbs, because in mixed suburbs, vertical extension of new houses is common. Crowded living condition is seen in the traditional suburbs with internal density of houses (i.e. more number of persons per room).

Residents, with their low income, are forced to live in this traditional suburbs where availability of facilities is also very low (Table 10.22). This type of suburbs have very few shops, bus services and metalled roads but some of the maujas (i.e. Ranip), due to industrialisation have some other amenities like police station, bank, etc., which are not present in mixed suburbs. So, the traditional suburbs have high score in amenities than the mixed type of suburbs. But private land agents, who have developed suburban areas are not interested in the development of traditional suburbs in terms of shops, approach roads, due to the low land value of this area. So, the traditional suburbs are gradually decaying while modern suburbs with low density and with facilities are flourishing into a good residential area. Most of the people of the suburbs like to live in the modern suburbs.The modern suburbs create a good image within the suburban population. So, the people who cannot afford to live in the modern suburbs but have the desire to live there, feel frustrated. Some of them like to live in mixed suburbs which are closer in quality to modern suburbs. Due to this reason, areas adjacent to modern suburbs have started growing very fast. In those areas people with relatively lower economic status have started living in a crowded manner to compensate with the higher land and rental value. Private house builders also take this as an opportunity and make houses in all available areas, in a crowded manner. Most of these housing societies are far from the main road and without any proper approach road. In the construction work of this type of house starts increasing in modern and mixed suburbs, then physical condition of these suburbs will also deteriorate in near future.

Fig 10.53

RESIDENTIAL PATTERN OF THE SUBUR
SOCIO - DEMOGRAPHIC MODEL

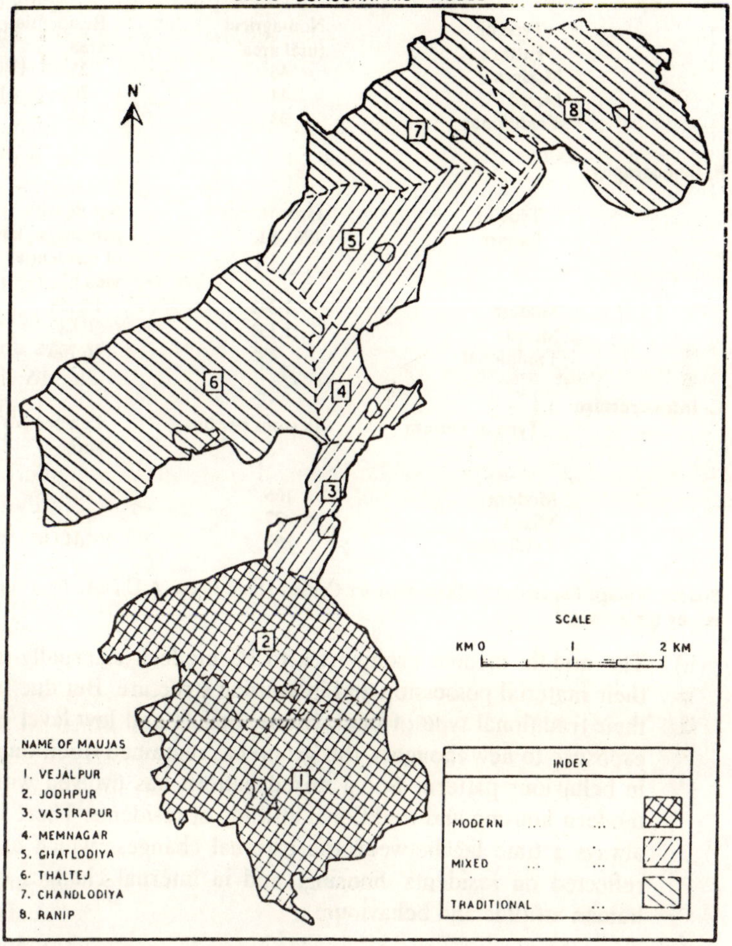

The following observations are noticed with the help of this model:

(i) The richer section of population is more exposed to the outer world due to their high income - hence they have modern attitude and behaviour in their life style as well as housing structure;

TABLE 10.22: **Distrbution of morphological factors in areas of different residential pattern, 1981**

A. Landuse (% to total area)

Type of Pattern	Non-agricultural area	Residential area
Modern	46	25
Mixed	44	20
Traditional	38	14

B. Demography

Type of Pattern	Growth rate 1961-81	Net density persons/sq.km of residential area
Modern	1,342	14,488
Mixed	1,151	20,687
Traditional	495	16,100

C. Infrastructure

Type of Pettern	Persons per bus	Persons per shop
Modern	166	166
Mixed	77	132
Traditional	257	764

Source: Village Panchayat Offices, District Census Hand Book, A.U.D.A., field survey by the author.

(ii) The middle income group tries to be modern externally in their material possession and in housing structure. But due to their traditional type of interactions, beliefs and low level of exposure to new thoughts, this group has remained traditional in behaviour pattern. So, a mixed pattern has evolved with modern housing and traditional attitude of residents. There is always a time lag between the external changes, which are reflected on residents' housing, and in internal changes of human attitude and behaviour;

(iii) The traditional group of residents have very little interaction with the modern world in terms of life style, housing condition and in behaviour pattern. People are still in a closed type of system and there is no clash between internal and external factors of life.

The conflict of philosophy of modernity and traditionality is more in the transitional zone than in the modern or traditional pattern of social organisation.

It is also observed in this model that the growth of the study area is very unsystamatic. Though the government has taken up many measures to control the suburban development but most of them have failed at certain level. A number of programmes is needed to have systematic suburban growth. The Urban Development Authority must provide equal amenities in all parts of the suburbs. The government must keep a programme of minimum requirement in suburban houses especially in the core villages which may become urban slums in future.

The 'Socio-Demographic Model' proposed in this study can be useful in analysing other Indian cities with some modifications. For a better understanding of the residential pattern and classification of the suburb, it is necessary to study this model at two points of time (at least five years apart). The study in two points of time will be helpful to have a picture in the trend of the pattern. the main drawback of this model is that it is static. It does not deal with constantly changing variables like migration of residents and housing market, i.e. buying and selling of the houses.

The model will certainly contribute towards the studies of pattern of suburbs especially in terms of total typology. A comparison of pattern between socio-economic status of residents, their behaviour and life style and their housing characteristics can bring out a clear picture in such a comprehensive manner that is useful in micro-level planning of suburban areas or any other areas where this model can be applied.

ELEVEN

CONCLUSIONS

1. Main findings

An examination of data and an experience of field work of various aspects of the suburbs of Ahmedabad city have revealed many interesting facts of the process of development of the suburbs. This includes changes in landuse, land value and population, infrastructural facilities, residents' socio-economic status, their behaviour and life style, interaction pattern and housing characteristics. The major findings can be outlined as follows:-

(a) This study has formulated a generalised model of step-wise migration of suburban population. This model shows that the suburban population have gone through different stages of migration within Ahmedabad city and study area before settling in their present location. Most of the people of the present study area are from other parts of Gujarat, different parts of Ahmedabad city and a few of them are from the core city of Ahmedabad and other parts of India.

(b) In the study it has been found that all core villages have traditional socio-economic status, behaviour and life style and housing characteristics, but housing societies in most of the maujas have modern socio-economic status but relatively traditional behaviour and life style.

(c) This study has also formulated a model, i.e 'Socio-Demographic Model' which has been represented by a three dimensional rectangular parallelopiped and depicts the residential pattern of the study area. A suburban typology has been evolved with the help of the model in terms of household and housing. Ranip, Chandlodiya and Thaltej situated in the extreme north, north-west and west have traditional pattern, whereas maujas Vejalpur and Jodhpur in south have modern pattern. Vastrapur, Memnagar and Ghatlodiya situated in the central part have mixed pattern. So this model has formulated a slope of modernity in the residential pattern from south to north. This model implies that behaviour and life style do not change as fast as socio-economic status or housing characteristics. It means people take longer time to change their behaviour and life sytle but external factors like housing, education, income can be changed faster. This model also implies that the conflict of philosophy of modernity and traditionality is more in areas where mixed type of residential pattern is present, i.e. transitional zone, then in the areas of modern or traditional pattern of social organisation.

(d) This study has also formulated slopes of various morphological factors from south to north. The northern side with low accessibility and connectivity has low residential area with low land value. The growth rate of population is low here, but the density of population per room is high and the level of infrastructural facilities is low. This side is far from the city-centre and have industrial landuse with bluecollar workers.

(e) The southern side with higher level of accessibility and connectivity has high residential area with high land value. The growth of population is high and the density of population per room is low. The level of infrastructural facilities is high. Here, the suburb is nearer to the city-centre and has people of high socio-economic status in the core villages as well as in the housing societies.

(f) The central portion of the study area, which has mixed type of residential pattern, also has transitional nature in various morphological factors. One mauja of this portion, i.e. Ghatlodiya, has high residential area with high growth of population and the level of infrastructural facilities are also high than other two maujas, i.e. Vastrapur and Memnagar of this area. But Ghatlodiya has high concentration of persons per room, low land value and high distance from the city-centre than Vastrapur and Memnagar.

Problems of residential suburbs

The rapid sprawl of the modern city outside its political boundaries possesses several problems.

(a) *Administrative problem*

Suburbs consist of a number of villages or maujas which are administered by different village panchayats. So the growth and development of a village depends mainly on the desire and outlook of local administrator. Due to the lack of a single administrative authority in the suburb, it is not growing in a continuous, uniform manner. Though in some places the development of infrastructure is taken care of, it may not be so at other places.

(b) *Problem related to landuse*

Due to the conversion of agricultural land, a large portion of fertile agricultural land is also converted to non-agricultural use. The loss of agricultural land in suburbs has increased the cost of vegetables due to the shortage of supply and extra transportation charges for carrying them from a longer distance to the city-centre.

Due to the mixed landuse pattern, i.e. agricultural and non-agricultural, the fertility of agricultural land has decreased, which on the other hand, has caused low productivity of agricultural products in suburbs. A large amount of agricultural land is left vacant in the suburbs which are purchased by speculators at the initial stage. This land speculation creates artificial scarcity of land in the suburbs and it also increases the price of land exorbitantly.

Though the Government has introduced the Urban Land Ceiling Act in 1976, it has no effect on the study area. According to this Act, conversion of agricultural land within 5 km from the periphery of city is prohibited, but this Act has failed to stop speculation in suburban land market. Speculation has started beyond 5 km leaving large area undeveloped between the city limit and the suburb.

(c) *Infrastructure*

The residential suburbs face various infrastructural problems related to housing, transportation, shopping, medical, banking, education and other services. These problems are ignored by municipal authorities as the region is outside the limit of Municipal Corporation and the residents do not pay any municipal taxes. Rural authorities also ignore these problems as the development of suburb is beneficial for city dwellers only.

Considering the housing facilities in the suburb, a number of problems become apparent, such as those of water supply, sewerage and electricity, as this belt is beyond the municipal limit.

Roads are generally in a poor condition in suburbs. There may be only one or two metalled roads which serves as main thoroughfares. Other roads of the suburbs are unmetalled and sometimes they are brick built or semi-metalled. Most of them get water-logged during the rainy season. The transport problem is also acute due to the lack of sufficient cheap public transport facilities. Frequency of buses are very low. Thus the suburban dwellers, who have high interaction with the city area for work, education, shopping, recreation, etc., commute in buses with great difficulty. In most of the areas of the residential suburbs buses generally run through one or two main roads. So it is very difficult for people who are living in the interior areas, to catch buses mainly in summer and rainy reasons. At times of emergency residents depend on autorickshaws which is the only alternative transport service available in suburbs. But the fare on this mode is high and are not easily available. Drivers of autorickshaws sometimes charge more to go to the suburban areas.

Medical facilities too are not adequate in the suburbs. In most of the places there is no public hospital. There may be one or two private nursing-homes present in the suburb. The charge of these nursing-homes is generally very high. There are few qualified doctors who are reluctant to go the suburbs leaving the city-centre. They think that suburban people can go to city for treatment but city dwellers will not go to suburbs for medical services. The same applies to chemists and druggists. Medical shops are generally estabilished near municipal boundary to capture customers from the suburbs and the city. Thus the conditions deteriorates in areas far from the city boundary.

There are also insufficient number of other shops in the suburbs. Cloth, utensils, ornaments and other luxury goods are purchased in the city but daily purchase of food items and consumer goods should be available in the suburb itself. As the number of shops are very few in comparison to the number of customers, the prices are very high. Sometimes even ordinary goods are not available in the suburb, and people have to cover a long distance to purchase them. The problems faced in times of need are enormous.

So far as education is concerned, there are very few high schools and colleges. For colleges and university education, students have to commute to the city, but in the case of high school and good primary school also there is scarcity. There may be some primary schools but they are of very low standard.

Lack of proper recreation facilities like park, cinema, restaurant etc. create many problems. New suburban residents are urbanised in nature, so they are used to travel to city for recreation mainly on holidays. Due to the lack of sufficient public transport system, there is generally great crowd in the city going buses during holidays.

The core villages of suburbs like other villages of India are devoid of any basic amenities like sewerage, sanitation, etc. and these are posing a contrast in comparison with newly developed areas of suburbs. Gradually these core villages will develop into urban slums when the density of population will increase and when these suburbs will be engulfed within the city limit.

(d) *Social aspect*

Due to the presence of people of various background, social problems are also present in the suburb. The original residents of the core villages have their social interaction within the villages while suburban people who are from the city core have their attachment very close with the city-centre, socially as well as economically. Inmigrated people also make their own group by language etc. in various parts of the city or within the suburb. But none of the suburban population mix with the villages. This type of interaction creates a dual character of suburbs and this dual character hampers the growth of understanding based on cultural diffusion and participation of all residents in development programmes is not possible.

3. Policy Implication

It is obvious from the study that there is a disparity within the study area in terms of various physical and behavioural aspects. This is bound to happen as the process of growth and development of suburbs are not properly planned. The government has taken up the challenge to control the haphazard growth of the city. In order to control this growth, Urban Land Ceiling Act was introduced in 1976 which stopped the transaction of land within 5 km from the periphery of the city. The government also introduced Urban Development Plan which will control the landuse pattern of the city and suburbs. But the main problem is that all these Acts were introduced in 1976, but transaction of land is still going on in the suburb. The government plans have changed from time to time. For example, the Green Belt, an area of 2 km around the city periphery was introduced in the seventies to check the urban growth but the government has abolished that idea in June 1983, and has kept it for the use of building houses for various government and semi-government organisations.

So, a number of rigid action programmes is needed which should be implemented in a comprehensive manner to have a well-planned residential suburb:

(i) All maujas which have potential for development must be administered by a single authority like suburban corporation to have equal distribution of facilities all over the area;
(ii) The government must implement all the landuse Act in a rigid manner;
(iii) All the future government urban land must be controlled by a government department which will sell land at the price to be fixed by the government;
(iv) Administrative authority must keep an eye on the growth of the suburb. The development of the suburb must be done in

various phases in relation to the distance from the city limit. The administrative authority must provide good shopping, education, medical, recreation and other facilities in the sub-urban area. It will decrease the pressure of amenities on the city-centre. Each suburban house must have minimum facility fixed by the authority. Along with the new residential area, authority must provide facilites to minimise disparity between housing societies and core villages and also to control the development of urban slums in core villages in the near future.

(v) Both the government and the residents should come forward to implement all plans rigidly. Awareness and perception of all residents in suburban planning are essential elements of long term progress by which the whole community will be benefited.

4. Scope for Future Study

There is enormous scope in the study of various aspects of suburbs. This study has focussed only on a few aspects. Under-standing of the city and suburbs will be useful if in future research is done on the suburb at various points of time. If it is studied at various points of time - socio-economic status and behaviour and life style of residents will give an idea of change in the human characteristics along with the change in the human physical characteristics of the suburb. Future study can also be done by comparing suburbs of various level cities. Detailed data are necessary at mauja level about various aspects of population. Similarly, detailed data are necessary about the core villages and housing societies separately to understand the problem of unsystemmatic pattern of suburbs so that future planning can be carried out in a meaningful manner.

APPENDICES

APPENDIX - 1

Ahmedabad Standard Urban Area (1971), Urban Components

I.	Ahmedabad City	VIII.	Naroda
II.	Bagfardos	IX.	Odhav
III.	Rakhiyal	X.	Ahmedabad Cantonment
IV.	Asarwa	XI.	Ranip
V.	Khokhara Mehmdabad	XII.	Sarkhej
VI.	Sardarnagar	XIII.	Nikol
VII.	Saijpur Bogha	XIV.	Hansol

Rural Components

1.	Ghatlodiya	27.	Hanspura
2.	Chandlodiya	28.	Muthiya
3.	Kotarpur	29.	Enasan
4.	Memnagar	30.	Bilasiya
5.	Vastrapur	31.	Kathwada
6.	Vejalpur	32.	Singarva
7.	Makarba	33.	Vastral
8.	Maktampur	34.	Ramol
9.	Ghodasar	35.	Vinzol
10.	Isanpur	36.	Vatva
11.	Shahwadi	37.	Fatewadi
12.	Gyaspur	38.	Gamdi
13.	Saijpur Gopalpur	39.	Geratpur
14.	Narol	40.	Ropda
15.	Khodiyar	41.	Sanathal
16.	Tragad	42.	Khoraj
17.	Chharodi	43.	Zundal
18.	Jagatpur	44.	Amiyapur
19.	Kali	45.	Bhat
20.	Chenpur	46.	Chandkheda
21.	Gota	47.	Motera
22.	Sola	48.	Koteshwar
23.	Thaltej	49.	Chiloda
24.	Bopal	50.	Sugad
25.	Ambli	51.	Badrabad
26.	Bodakdev	52.	Okaf

Source: Census of India, Standard Urban Area, 1971.

APPENDIX - 2

Calculation of 'Suburbanisation Index'

Name of Maujas	Veri-ables	1 Distance of village core from municipal boundary (KM)	2 Distance of mauja boundary from city center (KM)	3 Regis-tered Housing Society 1971	4 A.M.T.S. Bus Route 1971	5 A.M.T.S Bus fre_quency
	Weigh-tage	8	7	6	5	4
1	Vejalpur	2.48	6.18	185	11	277
2	Ranip	1.80	8.30	65	5	116
3	Vastrapur	0	5.25	5	3	79
4	Memnagar	0.25	5.00	10	4	107
5	Ghatlodiya	0.65	6.75	20	5	166
6	Chandlodiya	1.00	6.75	15	1	38
7	Sarkhej	4.80	12.00	1	4	128
8	Thaltej	1.80	6.25	10	1	45
9	Kali	3.25	10.75	12	1	8
10	Sola	5.50	10.50	2	1	21
11	Ambli	4.75	9.40	0	2	37
12	Bodakdev	1.80	5.75	2	0	0
13	Gota	7.00	10.75	0	1	0
14	Makarba	4.80	12.00	1	0	0
15	Bopal	8.40	11.40	0	1	17
16	Chenpur	9.25	11.85	3	0	0
17	Fatewadi	6.40	16.20	1	0	0
18	Jagatpur	11.40	14.50	2	0	0
19	Tragad	12.50	16.75	1	0	0
20	Chharodi	11.80	15.30	0	0	0
21	Santhal	9.60	18.40	0	0	0
	Mean ($\bar{x}$)	5.20	10.48	16.00	1.91	46.86
	Standard Deviation σ	3.92	3.98	40.30	2.63	70.77

Contd.

APPENDIX - 2 (CONTD.)

	Name of Maujas / Variables	6 Growth rate of population 1961-71	7 Density of Population 1971	8 Non-agricultural worker 1971
	Weightage	3	2	1
1	Vejalpur	150.83	720.88	69.52
2	Ranip	99.42	1854.40	93.60
3	Vastrapur	177.21	1054.40	65.98
4	Memnagar	151.56	1334.08	75.55
5	Ghatlodiya	16.36	245.41	37.54
6	Chandlodiya	54.93	435.75	55.34
7	Sarkhej	35.97	1462.77	77.99
8	Thaltej	28.42	439.60	36.94
9	Kali	111.01	264.87	94.70
10	Sola	73.45	470.60	58.55
11	Ambli	26.45	489.60	25.31
12	Bodakdev	31.75	84.46	14.79
13	Gota	39.68	297.60	40.62
14	Makarba	50.37	314.70	53.16
15	Bopal	17.35	323.20	39.84
16	Chenpur	49.76	450.60	74.92
17	Fatewadi	40.27	242.27	13.46
18	Jagatpur	44.11	236.69	47.49
19	Tragad	22.14	249.44	34.82
20	Chharodi	84.02	254.09	41.51
21	Santhal	45.94	164.92	23.09
	Mean ($\bar{X}$)	70.81	537.18	50.71
	Standard Deviation σ	55.05	480.84	23.04

Contd.

APPENDIX - 2 (CONTD.)

	Name of Maujas	Total points obtained	Suburbanisation Index	Pattern of suburbs
1	Vejalpur	556	77.22	U
2	Ranip	453	62.92	U
3	Vastrapur	442	61.38	U
4	Memnagar	437	60.69	U
5	Ghatlodiya	420	58.53	MU
6	Chandlodiya	379	52.64	MU
7	Sarkhej	378	52.50	MU
8	Thaltej	372	51.67	MU
9	Kali	356	49.44	MU
10	Sola	332	46.11	MU
11	Ambli	305	42.36	MR
12	Bodakdev	287	39.86	MR
13	Gota	263	36.53	MR
14	Makarba	261	36.25	MR
15	Bopal	258	35.83	MR
16	Chenpur	249	34.58	MR
17	Fatewadi	221	30.69	MR
18	Jagatpur	203	28.19	R
19	Tragad	192	26.67	R
20	Chharodi	155	21.53	R
21	Santhal	148	20.56	R
	Mean ($\bar{x}$)		44.10	
	Standard Deviation σ		14.74	

Calculation of points of each variable:

Mean = 10 points;

$\pm$ 1 point for $\pm \frac{1}{2}$ standard deviation
$\pm$ 1 point for $^{-}\frac{1}{+2}$ standard deviation,
in case of distance;
Maximum point of a variable is 20;

$$\text{Suburbanisation Index} = \frac{\text{Total point obtained} \times 100}{\text{Total maximum point}}$$

Total maximum point = (Weightage of a variable x Maximum point, i.e. 20)

U = Urban
($> \bar{x} + \sigma$)

MU = Moderately Urban
($\bar{x}$ to $\bar{x} + \sigma$)

MR = Moderately Rural
($\bar{x} - \sigma$ to $\bar{x}$)

R = Rural
($< \bar{x} - \sigma$)

Source: District Co-operative society office, A. M. T. S. office and District census hand book.

APPENDIX - 3

Calculation of 'Residential Index'

		1	2	3	4	5
Name of Maujas	Veri-ables	Distance of villa-ge core from city center (KM)	Percen-tage of residen-tial area 1981	No. of-built-up housing Socities 1981	A. M. T. S. bus routes 1981	A. M. T. S. bus frequency 1981
	Weigh-tage	5	4	3	2	1
1	Vejalpur	6.45	52.40	115	6	194
2	Jodhpur	5.90	55.00	46	5	83
3	Ghatlodiya	6.75	51.10	34	5	166
4	Memnagar	5.00	45.20	17	4	107
5	Ranip	8.30	34.30	73	5	116
6	Vastrapur	5.25	30.90	8	3	79
7	Thaltej	6.25	37.00	6	1	45
8	Chandlodiya	6.75	53.10	26	1	38
9	Sarkhej	12.00	19.00	1	4	128
10	Kali	10.75	27.10	8	1	8
11	Sola	10.50	21.90	1	1	21
	Mean ($\bar{x}$)	7.67	38.8	30.45	3.27	89.55
	Standard Deviation σ	2.41	13.2	34.05	1.86	56.72

Contd.

APPENDIX - 3 (CONTD.)

	Name of Maujas	Total points obtained	Residential Index	Pattern of suburbs
1	Vejalpur	122	81.33	PR
2	Jodhpur	107	71.33	PR
3	Ghatlodiya	98	65.33	PR
4	Memnagar	98	65.33	PR
5	Ranip	88	58.67	SR
6	Vastrapur	77	51.33	SR
7	Thaltej	66	44.00	SR
8	Chandlodiya	65	43.33	SR
9	Sarkhej	41	27.33	NR
10	Kali	39	26.00	NR
11	Sola	29	19.33	NR
	Mean ($\bar{x}$)		50.30	
	Standard Deviation σ		20.23	

Calculation of points of each variable:

Mean = 5 points;

$\pm$ 1 points for $\pm\frac{1}{2}$ standaed deviation

$\pm$ 1 points for $\frac{-1}{+2}$ standard deviation, in case of distance;

Maximum point of a variable is 10;

$$\text{Residental Index} = \frac{\text{Total point obtained} \times 100;}{\text{Total maximum point}}$$

Total maximum point = (Weightage of a variable x Maximum point, i.e. 10);

Type of suburbs:

PR = Purely Residential $(> \bar{x} + \frac{1}{2}\sigma)$

SR = Semi Residential $(\bar{x} \pm \frac{1}{2}\sigma)$

NR = Non Residential $(< \bar{x} - \frac{1}{2}\sigma)$

Source: Village Panchayat Office and A. M. T. S. Office.

APPENDIX - 4

The following variables have been used for the calculation of the 'Modernity Index' (in percentage):

A - Information regarding head of the family

Table 1	:	Place of origin
Table 2	:	Level of education
Table 3	:	Level of occupation

B - Information regarding total family

Table 4	:	Type of family
Table 5	:	Family members
Table 6	:	Relation with the head of the family
Table 7	:	Child population
Table 8	:	Young unmarried family members
Table 9	:	Level of education of family members
Table 10	:	Level of occupation of family members

C - Female participation

Table 11	:	Number of female students
Table 12	:	Level of occupation of female members in the family
Table 13	:	Female literacy

D - Socio-cultural-economic structure

Table 14	:	Food habit
Table 15	:	Religion
Table 16	:	Mother tongue
Table 17	:	Caste
Table 18	:	Total monthly income

E - Possession and use of modern goods

Table 19	:	Possession of modern goods
Table 20	:	Use of modern goods

F - Reading and recreation habit

Table 21	:	Persons keeping newspaper
Table 22	:	Language of newspaper
Table 23	:	Persons keeping magazine
Table 24	:	Type of magazine

F - Reading and recreation habit (Contd.)

Table 25	:	Language of magazine
Table 26	:	Number of library members
Table 27	:	Type of library
Table 28	:	Number of club members
Table 29	:	Type of club

G - Behaviour in raising of children

Table 30	:	Expected degree for male child
Table 31	:	Expected degree for female child
Table 32	:	Criteria for the mate selection in children's
Table 33	:	marriage, first and second preference.
Table 34	:	Number of diseases children below 10 years had in the last 5 years.
Table 35	:	Type of treatment given to a child below 10 years.
Table 36	:	Number of children whom vaccine has been applied.

H - Attitude towards banks and savings

Table 37	:	Number of people having account in savings bank.
Table 38	:	Number of people having shares and certificates.
Table 39	:	Number of shares and certificates.
Table 40	:	Number of people using safe deposit vault.

I - Knowledge about city

Table 41	:	Idea about city limit
Table 42	:	Idea about suburban limit

J - Structure of the house

Table 43	:	Material of the wall
Table 44	:	Material of the room
Table 45	:	Number of rooms
Table 46	:	Built-up area in square yard

K - Facilities available in the house

Table 47	:	Number of households having separate facility of electricity, water tap, bath room, kitchen and toilet.

L - Value of the house

Table 48	:	Rent in rupees (monthly)
Table 49	:	Selling value (in thousand rupees)

M - Internal density of the house

Table 50	:	Persons per room
Table 51	:	Person per square yard of built-up area.

Source: Filed Survey by the author.

APPENDIX - 4 : 'A' Information Regarding Head of the Family

TABLE 1 : **Place of Origin**

Name of Maujas	Type of residence	Vari-ables	Same Mauja (a)	Core City (b)	Other parts of Gujarat (c)	Anywhere in India (d)	Tota (%)
1 Vejalpur	i)	Core village	90	--	10	--	100
	ii)	Housing Soc.	1	23	60	16	100
2 Jodhpur	i)	Core village	87	--	13	--	100
	ii)	Housing Soc.	8	12	51	29	100
3 Vastrapur	i)	Core village	68	--	26	6	100
	ii)	Housing Soc.	--	8	65	27	100
4 Memnagar	i)	Core village	84	--	16	--	100
	ii)	Housing Soc.	5	15	73	7	100
5 Ghatlodiya	i)	Core village	86	--	7	7	100
	ii)	Housing Soc.	--	12	68	20	100
6 Thaltej	i)	Core village	87	--	13	--	100
	ii)	Housing Soc.	--	18	72	10	100
7 Chandlodiya	i)	Core village	83	--	17	--	100
	ii)	Housing Soc.	--	11	80	9	100
8 Ranip	i)	Core village	56	20	20	4	100
	ii)	Housing Soc.	1	4	84	11	100
9 Total Study Area	i)	Core village	76	1	18	5	100
	ii)	Housing Soc.	2	14	67	17	100

Contd.

APPENDIX - 4 : 'A' (CONTD.)

TABLE 2 : Level of Education

Name of Maujas		Type of residence	Illi-terate (a)	Primary (b)	Secon-dary (c)	Higher Secon-dary (d)	Graduate (e)	M.A., Law Engineer, Doctor (f)	Ph.D and above (g)	Total (%)
1 Vejalpur	i)	Core village	30	64	3	3	--	--	--	100
	ii)	Housing Soc.	7	16	10	17	25	17	8	100
2 Jodhpur	i)	Core village	7	60	20	6	7	--	--	100
	ii)	Housing Soc.	--	8	6	22	25	33	6	100
3 Vastrapur	i)	Core village	37	53	5	5	--	--	--	100
	ii)	Housing Soc.	--	--	15	8	38	31	8	100
4 Memnagar	i)	Core village	18	50	--	32	--	--	--	100
	ii)	Housing Soc.	7	15	27	12	15	24	--	100
5 Ghatlodiya	i)	Core village	6	94	--	--	--	--	--	100
	ii)	Housing Soc.	3	13	27	14	26	16	11	100
6 Thaltej	i)	Core village	41	53	3	3	--	--	--	100
	ii)	Housing Soc.	--	12	32	4	28	24	--	100
7 Chandlodiya	i)	Core village	29	47	24	--	--	--	--	100
	ii)	Housing Soc.	4	41	27	12	10	6	--	100
8 Ranip	i)	Core village	16	64	12	8	--	--	--	100
	ii)	Housing Soc.	8	19	25	13	26	9	--	100
9 Total Study Area	i)	Core village	18	68	8	6	--	--	--	100
	ii)	Housing Soc.	1	17	19	19	24	19	1	100

Contd.

APPENDIX - 4 : 'A' (CONTD.)

TABLE 3 : Level of Occupation

Name of Maujas	Type of residence	Variables	Primary occupation, occupation of own caste (a)	Mill worker and peon (b)	Clerk (c)	Business man (d)	Officer Profession (e)	Total (%)
1 Vejalpur	i)	Core village	87	--	--	13	--	100
	ii)	Housing Soc.	12	6	25	23	34	100
2 Jodhpur	i)	Core village	79	--	13	8	--	100
	ii)	Housing Soc.	11	2	22	23	42	100
3 Vastrapur	i)	Core village	48	5	21	26	--	100
	ii)	Housing Soc.	18	7	30	15	30	100
4 Memnagar	i)	Core village	68	8	8	8	8	100
	ii)	Housing Soc.	28	2	27	18	25	100
5 Ghatlodiya	i)	Core village	63	--	6	31	--	100
	ii)	Housing Soc.	14	7	38	21	20	100
6 Thaltej	i)	Core village	76	7	--	17	--	100
	ii)	Housing Soc.	24	4	28	24	20	100
7 Chandlodiya	i)	Core village	82	--	18	--	--	100
	ii)	Housing Soc.	25	1	28	39	7	100
8 Ranip	i)	Core village	60	4	4	32	--	100
	ii)	Housing Soc.	25	7	25	33	10	100
9 Total Study Area	i)	Core village	73	2	5	18	2	100
	ii)	Housing Soc.	17	6	30	24	23	100

APPENDIX - 4 : 'B' - Information Regarding Total Family

TABLE 4 : **Type of Family** TABLE 5 : **Family Members**

Name of Maujas		Type of residence	Joint (a)	Nuclear (b)	Total (%)	More than 9 (a)	7 - 9 (b)	4 - 6 (c)	Less than 4 (d)	Total (%)
1 Vejalpur	i)	Core village	80	20	100	40	37	20	3	100
	ii)	Housing Soc.	49	51	100	6	62	19	13	100
2 Jodhpur	i)	Core village	87	13	100	13	60	27	--	100
	ii)	Housing Soc.	32	68	100	--	9	68	23	100
3 Vastrapur	i)	Core village	68	32	100	27	52	16	5	100
	ii)	Housing Soc.	28	72	100	--	16	46	38	100
4 Memnagar	i)	Core village	83	17	100	27	32	33	8	100
	ii)	Housing Soc.	37	63	100	7	5	71	17	100
5 Ghatlodiya	i)	Core village	64	36	100	25	50	13	12	100
	ii)	Housing Soc.	36	64	100	4	14	67	15	100
6 Thaltej	i)	Core village	84	16	100	30	43	27	--	100
	ii)	Housing Soc.	38	62	100	12	--	68	20	100
7 Chandlodiya	i)	Core village	50	50	100	22	33	28	17	100
	ii)	Housing Soc.	49	51	100	10	14	59	17	100
8 Ranip	i)	Core village	60	40	100	20	28	48	4	100
	ii)	Housing Soc.	41	59	100	6	20	55	19	100
9 Total Study Area	i)	Core village	64	36	100	25	38	30	7	100
	ii)	Housing Soc.	42	58	100	5	28	50	17	100

Contd.

APPENDIX - 4 : 'B' (CONTD.)

TABLE 6 : **Relation with the Head of the Family**

Name of Maujas		Type of residence	Variables	Very distant relation (a)	Cousin nephew (b)	Unmarried neice, (c)	Married brother (d)	Married great grandson (e)	Unmarried great grand son (f)	Married grandson (g)
1 Vejalpur	i)	Core village		1	1	1	—	1	4	5
	ii)	Housing Soc.		1	1	3	2	—	1	1
2 Jodhpur	i)	Core village		-	-	7	2	-	-	-
	ii)	Housing Soc.		-	-	-	-	-	1	1
3 Vastrapur	i)	Core village		-	-	-	-	-	-	-
	ii)	Housing Soc.		-	-	-	-	-	-	-
4 Memnagar	i)	Core village		-	-	-	-	-	3	3
	ii)	Housing Soc.		-	-	-	-	-	-	-
5 Ghatlodiya	i)	Core village		-	-	-	-	-	-	-
	ii)	Housing Soc.		-	-	-	1	-	-	-
6 Thaltej	i)	Core village		-	-	-	-	-	1	4
	ii)	Housing Soc.		-	-	-	-	-	-	-
7 Chandlodiya	i)	Core village		-	-	2	-	-	-	-
	ii)	Housing Soc.		-	-	-	-	-	—	2
8 Ranip	i)	Core village		-	-	-	-	-	-	4
	ii)	Housing Soc.		-	-	1	-	-	3	-
9 Total Study Area	i)	Core village		1	1	1	1	1	1	1
	ii)	Housing Soc.		1	1	1	1	1	1	1

Contd.

APPENDIX - 4 : 'B' (CONTD.)

Name of Maujas		Type of residence (Variables)	Unmarried brothers & sisters (h)	Unmarried grand children in-law (i)	Married son and daughter- (j)	Widowed mother (k)	Unmarried child (l)	Wife (m)	Total (%)
1 Vejalpur	i)	Core village	1	21	25	2	27	11	100
	ii)	Housing Soc.	1	10	11	2	48	19	100
2 Jodhpur	i)	Core village	4	22	13	3	36	13	100
	ii)	Housing Soc.	1	6	13	2	51	25	100
3 Vastrapur	i)	Core village	-	19	29	1	38	13	100
	ii)	Housing Soc.	-	2	13	2	55	28	100
4 Memnagar	i)	Core village	-	25	20	4	32	13	100
	ii)	Housing Soc.	1	13	20	-	45	21	100
5 Ghatlodiya	i)	Core village	4	14	37	2	26	17	100
	ii)	Housing Soc.	2	7	10	2	54	24	100
6 Thaltej	i)	Core village	1	16	19	1	45	13	100
	ii)	Housing Soc.	2	10	16	-	47	25	100
7 Chandlodiya	i)	Core village	1	18	25	3	38	13	100
	ii)	Housing Soc.	2	13	16	2	46	19	100
8 Ranip	i)	Core village	5	15	17	3	41	15	100
	ii)	Housing Soc.	2	7	13	2	52	21	100
9 Total Study Area	i)	Core village	1	18	23	2	36	13	100

Contd.

APPENDIX - 4 : 'B' (CONTD.)

Name of Maujas		Type of residence	TABLE 7: **Children** Percentage of children below 5 yrs. in age group 11-20	TABLE 8 : **Single Population** Percentage of single population	TABLE 9 : **Level of Educatopm of Family Members** Illiterate (a)	No. of Sch. primary students (b)	Secondary primary students (c)	Secondary High school students (d)
1 Vejalpur	i)	Core village	24	67	27	27	37	4
	ii)	Housing Soc.	17	97	11	22	27	9
2 Jodhpur	i)	Core village	14	69	31	23	31	9
	ii)	Housing Soc.	11	100	6	18	23	9
3 Vastrapur	i)	Core village	33	60	46	22	28	3
	ii)	Housing Soc.	9	100	11	21	25	3
4 Memnagar	i)	Core village	31	73	33	18	33	13
	ii)	Housing soc.	23	97	14	15	31	14
5 Ghatlodiya	i)	Core village	23	75	44	27	15	10
	ii)	Housing Soc.	21	98	16	21	28	12
6 Thaltej	i)	Core village	35	60	41	33	15	10
	ii)	Housing Soc.	28	90	15	18	30	12
7 Chandlodiya	i)	Core village	28	83	36	26	31	7
	ii)	Housing Soc.	26	96	15	21	35	20
8 Ranip	i)	Core village	38	78	49	23	19	8
	ii)	Housing Soc.	23	91	13	25	31	13
9 Total Study Area	i)	Core village	30	71	38	23	28	6
	ii)	Housing Soc.	23	96	12	21	28	11

Contd.

APPENDIX - 4 : 'B' (CONTD.)

TABLE 9: Level of Education of Family Members (CONTD.)

Name of Maujas		Type of residence	Variables	High School Vocational trg. student (e)	College Student (f)	Graduate (g)	Engineering Medicine (h)	Lawyer Engineer (i)	Ph.D & above (j)	Total (%)
1 Vejalpur	i)	Core village		2	1	1	1	-	-	100
	ii)	Housing Soc.		5	7	13	3	2	1	100
2 Jodhpur	i)	Core village		1	2	2	-	1	-	100
	ii)	Housing Soc.		5	8	21	6	3	1	100
3 Vastrapur	i)	Core village		1	1	-	-	-	-	100
	ii)	Housing Soc.		3	10	8	6	10	3	100
4 Memnagar	i)	Core village		2	-	-	1	-	-	100
	ii)	Housing Soc.		5	6	11	3	1	-	100
5 Ghatlodiya	i)	Core village		2	-	1	-	1	-	100
	ii)	Housing Soc.		8	5	6	1	2	1	100
6 Thaltej	i)	Core village		1	-	-	-	-	-	100
	ii)	Housing Soc.		6	1	13	1	4	-	100
7 Chandlodiya	i)	Core village		-	-	-	-	-	-	100
	ii)	Housing Soc.		4	1	2	1	1	-	100
8 Ranip	i)	Core village		-	1	-	-	-	-	100
	ii)	Housing Soc.		4	5	7	1	1	-	100
9 Total Study Area	i)	Core village		1	1	1	1	1	-	100
	ii)	Housing Soc.		6	6	10	2	2	2	100

Contd.

APPENDIX - 4 : 'B' (CONTD.)

TABLE 10: **Level of Occupation of Family Members**

Name of Maujas		Type of residence / Variables	Agri-culture (a)	House wife retired unemployed etc. (b)	Student (c)	Mill worker & peon (d)	Clerk (e)	Busi-nessman (f)	Offi-cer profes-sion (g)	Total (%)
1 Vejalpur	i)	Core village	17	45	33	-	2	2	1	100
	ii)	Housing Soc.	-	38	44	1	8	4	5	100
2 Jodhpur	i)	Core village	30	36	26	-	7	-	1	100
	ii)	Housing Soc.	-	36	47	1	5	3	8	100
3 Vastrapur	i)	Core village	2	56	26	2	6	4	4	100
	ii)	Housing Soc.	-	49	41	-	2	-	8	100
4 Memnagar	i)	Core village	3	60	26	-	10	-	1	100
	ii)	Housing Soc.	-	42	44	-	8	2	4	100
5 Ghatlodiya	i)	Core village	-	50	26	2	8	4	-	100
	ii)	Housing Soc.	-	45	45	1	5	2	2	100
6 Thaltej	i)	Core village	14	68	17	-	-	1	-	100
	ii)	Housing Soc.	-	48	43	-	6	1	2	100
7 Chandlodiya	i)	Core village	6	61	29	-	4	-	-	100
	ii)	Housing Soc.	2	50	34	1	8	3	2	100
8 Ranip	i)	Core village	7	61	22	-	5	4	1	100
	ii)	Housing Soc.	-	43	45	1	6	3	2	100
9 Total Study Area	i)	Core village	5	58	28	2	4	2	1	100
	ii)	Housing Soc.	-	44	43	1	5	4	3	100

APPENDIX - 4 : 'C' - FEMALE PARTICIPATION

Name of Maujas		Type of residence	TABLE 11 Students: % of female students 5-20 age group	TABLE 12 Level of Occupation of Female Members: House wife unemployed (a)	Clerk (b)	Profes-sion (c)	Total (%)	TABLE 13 : Literacy: % of female literacy
1 Vejalpur	i)	Core village	63	100	-	-	100	64
	ii)	Housing Soc.	98	76	14	10	100	88
2 Jodhpur	i)	Core village	73	100	-	-	100	78
	ii)	Housing Soc.	100	75	13	12	100	100
3 Vastrapur	i)	Core village	45	100	-	-	100	36
	ii)	Housing Soc.	100	85	15	-	100	100
4 Memnagar	i)	Core village	65	100	-	-	100	33
	ii)	Housing Sco.	96	82	16	2	100	84
5 Ghatlodiya	i)	Core village	53	100	-	-	100	64
	ii)	Housing Soc.	95	94	3	3	100	83
6 Thaltej	i)	Core village	50	100	-	-	100	63
	ii)	Housing Soc.	91	89	7	4	100	90
7 Chandlodiya	i)	Core village	43	100	-	-	100	50
	ii)	Housing Soc.	70	100	-	-	100	80
8 Ranip	i)	Core village	48	100	-	-	100	42
	ii)	Housing Soc.	71	100	-	-	100	81
9 Total Study Area	i)	Core village	53	100	-	-	100	42
	ii)	Housing Soc.	94	86	8	6	100	88

Vari-ables

APPENDIX - 4 : 'D' - Socio-Cultural-Economic Structure

TABLE 14: **Food Habit** / TABLE 15: **Religion**

Name of Maujas	Type of residence	Variables	Vegetarian (a)	Only egg permitted rian (b)	Non-vegetarian (c)	Total (%)	Hindu (a)	Jain (b)	Muslim (c)	Others (d)	Total (%)
1 Vejalpur	i)	Core village	100	-	-	100	100	-	-	-	100
	ii)	Housing Soc.	64	11	25	100	74	3	20	3	100
2 Jodhpur	i)	Core village	100	-	-	100	100	-	-	-	100
	ii)	Housing Soc.	45	34	21	100	94	1	2	3	100
3 Vastrapur	i)	Core village	95	-	5	100	100	-	-	-	100
	ii)	Housing Soc.	54	18	30	100	92	8	-	-	100
4 Memnagar	i)	Core village	100	-	-	100	100	-	-	-	100
	ii)	Housing Soc.	70	20	10	100	95	5	-	-	100
5 Ghatlodiya	i)	Core village	100	-	-	100	100	-	-	-	100
	ii)	Housing Soc.	73	12	15	100	88	4	-	8	100
6 Thaltej	i)	Core village	100	-	-	100	100	-	-	-	100
	ii)	Housing Soc.	84	-	16	100	93	5	-	2	100
7 Chandlodiya	i)	Core village	94	-	6	100	95	-	5	-	100
	ii)	Housing Soc.	96	2	2	100	96	3	-	1	100
8 Ranip	i)	Core village	60	-	40	100	70	-	30	-	100
	ii)	Housing Soc.	92	3	5	100	95	2	1	2	100
9 Total Study Area	i)	Core village	93	-	7	100	93	-	7	-	100
	ii)	Housing Soc.	72	12	16	100	87	3	8	2	100

Contd.

APPENDIX - 4 : 'D' (CONTD.)

TABLE 16 : Mother Tongue

Name of Maujas	Variables / Type of residence	Gujarati (a)	Language of neighbouring state (b)	Hindi (c)	Other Indian Language (d)	Total (%)
1 Vejalpur	i) Core village	100	-	-	-	100
	ii) Housing Soc.	85	10	-	5	100
2 Jodhpur	i) Core village	100	-	-	-	100
	ii) Housing Soc.	75	5	4	16	100
3 Vastrapur	i) Core village	95	5	-	-	100
	ii) Housing Soc.	79	-	7	14	100
4 Memnagar	i) Core village	92	8	-	-	100
	ii) Housing Soc.	70	7	7	16	100
5 Ghatlodiya	i) Core village	88	12	-	-	100
	ii) Housing Soc.	80	5	6	9	100
6 Thaltej	i) Core village	100	-	-	-	100
	ii) Housing Soc.	80	-	8	12	100
7 Chandlodiya	i) Core village	100	-	-	-	100
	ii) Housing Soc.	93	3	3	1	100
8 Ranip	i) Core village	80	4	16	-	100
	ii) Housing Soc.	90	6	2	2	100
9 Total Study Area	i) Core village	95	3	2	-	100
	ii) Housing Soc.	84	6	3	7	100

Contd.

APPENDIX - 4 : 'D' (CONTD.)

TABLE 17 : **Caste**

Name of Maujas	Type of residence	Vari-ables	Patel (a)	Brahmin & Vaniya (b)	Thakore (c)	Harijan (d)	Others (e)	Total (%)
Vejalpur	i)	Core village	1	3	85	4	7	100
	ii)	Housing Soc.	24	18	23	12	23	100
2 Jodhpur	i)	Core village	63	3	24	2	8	100
	ii)	Housing Soc.	21	30	6	8	35	100
3 Vastrapur	i)	Core village	-	1	90	2	7	100
	ii)	Housing Soc.	24	32	-	1	43	100
4 Memnagar	i)	Core village	43	17	21	8	11	100
	ii)	Housing Soc.	21	16	18	4	41	100
5 Ghatlodiya	i)	Core village	81	6	3	1	9	100
	ii)	Housing Soc.	19	12	1	10	58	100
6 Thaltej	i)	Core village	1	1	94	1	3	100
	ii)	Housing Soc.	14	24	8	5	49	100
7 Chandlodiya	i)	Core village	48	6	12	10	24	100
	ii)	Housing Soc.	1	8	12	12	67	100
8 Ranip	i)	Core village	16	4	24	18	38	100
	ii)	Housing Soc.	5	6	6	14	69	100
9 Total Study Area	i)	Core village	42	4	46	5	3	100
	ii)	Housing Soc.	15	28	6	7	44	100

Contd.

APPENDIX - 4 : 'D' (CONTD.)

TABLE 18 : Total Monthly Income (in Rupees)

Name of Maujas	Type of residence	Variables	Less than 500 (a)	500-999 (b)	1000-1499 (c)	1500-1999 (d)	More than 1999 (e)	Total (%)
1 Vejalpur	i)	Core village	26	37	23	13	1	100
	ii)	Housing Soc.	3	40	33	16	8	100
2 Jodhpur	i)	Core village	13	53	20	7	7	100
	ii)	Housing Soc.	7	20	26	19	28	100
3 Vastrapur	i)	Core village	53	26	21	-	-	100
	ii)	Housing Soc.	-	15	46	23	16	100
4 Memnagar	i)	Core village	25	67	8	-	-	100
	ii)	Housing Soc.	-	38	34	14	14	100
5 Ghatlodiya	i)	Core village	26	64	10	-	-	100
	ii)	Housing Soc.	4	53	25	13	5	100
6 Thaltej	i)	Core village	20	40	27	13	-	100
	ii)	Housing Soc.	8	60	8	8	16	100
7 Chandlodiya	i)	Core village	28	72	-	-	-	100
	ii)	Housing Soc.	4	75	21	-	-	100
8 Ranip	i)	Core village	28	72	-	-	--	100
	ii)	Housing Soc.	-	60	29	6	5	100
9 Total Study Area	i)	Core village	30	50	14	5	1	100
	ii)	Housing Soc.	4	46	28	12	10	100

APPENDIX - 4 : 'E' - Possession and use of Modern Goods

TABLE 19: **Possession of Modern Goods**

Name of Maujas	Type of residence	Kerosene Stove (a)	Gas Stove (b)	Cooking range (c)	Pressure cooker (d)	Mixer (e)	Radio/ Trans-istor (f)	Record Player (g)	Tape Recor-der (h)
1 Vejalpur	i) Core village	57	7	-	-	-	53	-	7
	ii) Housing Soc.	99	56	3	94	48	100	5	39
2 Jodhpur	i) Core village	87	27	-	-	-	80	-	20
	ii) Housing Soc.	100	85	7	96	71	100	13	55
3 Vastrapur	i) Core village	79	-	-	-	-	73	-	-
	ii) Housing Soc.	100	77	-	100	92	100	8	77
4 Memnagar	i) Core village	83	8	-	-	-	89	-	-
	ii) Housing Soc.	98	51	2	73	22	98	12	24
5 Ghatlodiya	i) Core village	94	-	-	13	-	88	-	13
	ii) Housing Soc.	100	40	2	66	21	92	4	28
6 Thaltej	i) Core village	70	3	-	3	-	77	-	3
	ii) Housing Soc.	100	56	4	68	28	91	12	44
7 Chandlodiya	i) Core village	100	-	-	-	-	62	-	6
	ii) Housing Soc.	99	31	-	49	-	76	-	13
8 Ranip	i) Core village	100	-	-	4	-	68	-	-
	ii) Housing Soc.	100	55	-	55	9	89	2	21
9 Total Study Area	i) Core village	81	5	-	2	-	66	-	23
	ii) Housing Soc.	99	54	2	76	34	91	6	34

Contd.

APPENDIX - 4 : 'E' (CONTD.)

TABLE 19: **Possession of Modern Goods**

Name of Maujas	Type of residence	Television (i)	Clock (j)	Dining table (k)	Refrigerator (l)	Telephone (m)	Cycle (n)	Scooter (o)	Car (p)
1 Vejalpur	i) Core village	-	33	-	-	-	67	30	-
	ii) Housing Soc.	5	88	40	45	14	69	64	7
2 Jodhpur	i) Core village	-	47	-	-	-	87	20	-
	ii) Housing Soc.	9	95	58	65	26	72	68	15
3 Vastrapur	i) Core village	-	16	-	-	-	53	11	-
	ii) Housing Soc.	7	100	69	62	23	54	62	15
4 Memnagar	i) Core village	-	50	-	-	-	50	33	-
	ii) Housing Soc.	2	63	17	17	5	71	29	5
5 Ghatlodiya	i) Core village	6	31	-	-	-	69	19	-
	ii) Housing Soc.	3	80	21	21	21	85	23	7
6 Thaltej	i) Core village	-	33	-	-	-	73	13	-
	ii) Housing Soc.	-	88	36	4	12	60	52	-
7 Chandlodiya	i) Core village	-	17	-	-	-	83	-	-
	ii) Housing Soc.	-	51	3	4	-	89	8	-
8 Ranip	i) Core village	-	40	-	-	-	88	-	-
	ii) Housing Soc.	-	89	14	6	2	94	14	-
9 Total Study Area	i) Core village	-	33	-	-	-	72	15	-
	ii) Housing Soc.	4	83	30	30	11	78	41	6

Contd.

APPENDIX - 4 : 'E' (CONTD.)

TABLE 20: **Use of Modern Goods**

Name of Maujas	Type of residence		Vari-ables	Washing Soap soap (a)	Utensil cleansing (b)	Bathing soap (c)	Tooth paste (d)	Shampoo (e)	Safety razor (f)
1 Vejalpur	i)	Core village		100	-	63	7	-	30
	ii)	Housing Soc.		100	93	100	62	25	97
2 Jodhpur	i)	Core village		100	20	93	40	7	47
	ii)	Housing Soc.		100	100	100	79	40	98
3 Vastrapur	i)	Core village		95	5	90	11	-	47
	ii)	Housing Soc.		100	100	100	85	39	100
4 Memnagar	i)	Core village		100	8	100	42	-	100
	ii)	Housing Soc.		100	98	100	100	22	98
5 Ghatlodiya	i)	Core village		94	25	88	31	-	25
	ii)	Housing Soc.		100	94	99	84	21	97
6 Thaltej	i)	Core village		100	83	60.	7	-	20
	ii)	Housing Soc.		100	100	100	100	16	100
7 Chandlodiya	i)	Core village		94	-	67	22	-	39
	ii)	Housing Soc.		99	79	100	86	3	92
8 Ranip	i)	Core village		96	-	72	8	-	24
	ii)	Housing Soc.		100	95	100	69	10	99
9 Total Study Area	i)	Core village		98	6	75	19	12	36
	ii)	Housing Soc.		100	94	100	76	22	97

APPENDIX - 4 : 'F' - Reading & Recreation Habit

Name of Maujas		Type of residence	TABLE 21: Newspaper Vari-ables	TABLE 22: Language of Newspapers % of persons keeping newspaper	Mother tongue (a)	TABLE 23: Magazine Hindi (b)	TABLE 24: Type of magazine English (c)	% of persons keeping magazine	Religious (a)	Film & sports (b)
1 Vejalpur	i)	Core village		10	100	-	-	-	-	-
	ii)	Housing Soc.		88	89	-	14	27	17	19
2 Jodhpur	i)	Core village		47	100		-	-	-	-
	ii)	Housing Soc.		99	78	1	38	42	13	48
3 Vastrapur	i)	Core village		11	100	-	-	-	-	-
	ii)	Housing Soc.		85	72	-	-	44	29	57
4 Memnagar	i)	Core village		8	100	-	-	-	-	-
	ii)	Housing Soc.		83	100	-	21	36	25	15
5 Ghatlodiya	i)	Core village		6	100	-	-	-	-	-
	ii)	Housing Soc.		63	80	1	21	27	38	18
6 Thaltej	i)	Core village		3	100	-	-	-	-	-
	ii)	Housing Soc.		56	78	7	30	24	50	10
7 Chandlodiya	i)	Core village		6	100	-	-	-	-	-
	ii)	Housing Soc.		49	93	5	2	4	100	-
8 Ranip	i)	Core village		4	100	-	-	-	-	-
	ii)	Housing Soc.		75	93	2	5	23	64	12
9 Total Study	i)	Core village		8	100	-	-	-	-	-
Area	ii)	Housing Soc.		77	87	3	18	27	35	21

Contd.

APPENDIX - 4 : 'F' (CONTD.)

Name of Vari-ables		Type of residence	TABLE 24 : Type of Magazine (Contd.) General Maujas (c)	Educa- (d)	Others (e)	TABLE 25: Language of Magazine Mother tongue Language (a)	Any other Indian Language (b)	English (c)	TABLE 26: Library Members % of population Library members
1 Vejalpur	i)	Core village	-	-	-	-	-	-	-
	ii)	Housing Soc.	11	-	83	81	1	25	12
2 Jodhpur	i)	Core village	-	-	-	-	-	-	-
	ii)	Housing Soc.	7	9	95	82	7	33	33
3 Vastrapur	i)	Core village	-	-	-	-	-	-	-
	ii)	Housing Soc.	14	-	45	100	17	17	46
4 Memnagar	i)	Core village	-	-	-	-	-	-	-
	ii)	Housing soc.	-	1	30	100	13	26	17
5 Ghatlodiya	i)	Core village	-	-	-	-	-	-	-
	ii)	Housing Soc.	-	-	34	76	7	29	14
6 Thaltej	i)	Core village	-	-	-	-	-	-	-
	ii)	Housing Soc.	-	-	32	100	-	-	12
7 Chandlodiya	i)	Core village	-	-	-	-	-	-	-
	ii)	Housing Soc.	-	-	-	100	33	-	1
8 Ranip	i)	Core village	-	-	-	-	-	-	-
	ii)	Housing Soc.	6	-	-	86	3	10	7
9 Total Study Area	i)	Core village	-	-	-	-	-	-	-
	ii)	Housing Soc.	14	7	75	67	6	10	14

Contd.

APPENDIX - 4 : 'F' (CONTD.)

Name of Maujas		Type of residence	Vari-ables	TABLE 27: Type of Library — Circula-ting (a)	College & University (b)	Library having books of various types (c)	TABLE 28: Club Members — % of po-pulation as club members	TABLE 29: Type Of the Club — Local origi-nal (a)	Sports Club (b)	High status club (c)
1 Vejalpur	i)	Core village		-	-	-	-	-	-	-
	ii)	Housing Soc.		69	58	41	7	53	12	35
2 Jodhpur	i)	Core village		-	-	-	-	-	-	-
	ii)	Housing Soc.		51	23	85	13	29	7	64
3 Vastrapur	i)	Core village		-	-	-	-	-	-	-
	ii)	Housing Soc.		60	33	17	15	50	-	50
4 Memnagar	i)	Core village		-	-	-	-	-	-	-
	ii)	Housing Soc.		100	6	-	7	100	-	-
5 Ghatlodiya	i)	Core village		-	-	-	-	-	-	-
	ii)	Housing Soc.		100	8	-	1	100	-	-
6 Thaltej	i)	Core village		-	-	-	-	-	-	-
	ii)	Housing Soc.		98	12	-	-	-	-	-
7 Chandlodiya	i)	Core village		-	-	-	-	-	-	-
	ii)	Housing Soc.		100	-	-	-	-	-	-
8 Ranip	i)	Core village		-	-	-	-	-	-	-
	ii)	Housing Soc.		97	3	-	-	-	-	-
9 Total Study Area	i)	Core village		-	-	-	-	-	-	-
	ii)	Housing Soc.		57	50	34	5	49	8	43

APPENDIX - 4 : 'G' Behaviour in Raising of Children

TABLE 30: Expected Degree for Male Children

Name of Maujas		Type of residence	No expectation (a)	School Leaving (b)	Graduate (c)	Post graduate and equivalent (d)	Not fixed (as high as possible) (e)	Total (%)
1 Vejalpur	i)	Core village	-	76	24	-	-	100
	ii)	Housing Soc.	-	6	31	1	62	100
2 Jodhpur	i)	Core village	-	79	14	-	7	100
	ii)	Housing Soc.	-	-	18	-	82	100
3 Vastrapur	i)	Core village	9	80	11	-	-	100
	ii)	Housing Soc.	-	-	17	17	66	100
4 Memnagar	i)	Core village	10	80	10	-	-	100
	ii)	Housing Soc.	-	13	26	13	48	100
5 Ghatlodiya	i)	Core village	-	91	9	-	-	100
	ii)	Housing Soc.	1	15	52	6	26	100
6 Thaltej	i)	Core village	7	83	10	-	-	100
	ii)	Housing Soc.	-	9	48	14	29	100
7 Chandlodiya	i)	Core village	50	44	6	-	-	100
	ii)	Housing Soc.	-	54	39	-	7	100
8 Ranip	i)	Core village	34	65	1	-	-	100
	ii)	Housing Soc.	1	13	77	-	9	100
9 Total Study Area	i)	Core village	13	73	13	-	1	100
	ii)	Housing Soc.	1	13	43	3	40	100

Contd.

APPENDIX - 4 : 'G' (CONTD.)

TABLE 31: Expected Degree for Female Children

Name of Maujas		Type of residence	No expectation (a)	School Leaving (b)	Graduate (c)	Post graduate and equivalent (d)	Not fixed (as high as possible) (e)	Total (%)
1 Vejalpur	i)	Core village	-	85	15	-	-	100
	ii)	Housing Soc.	-	17	34	-	49	100
2 Jodhpur	i)	Core village	-	77	15	-	8	100
	ii)	Housing Soc.	-	-	21	5	74	100
3 Vastrapur	i)	Core village	50	50	-	-	-	100
	ii)	Housing soc.	-	-	40	20	40	100
4 Memnagar	i)	Core village	45	33	22	-	-	100
	ii)	Housing soc.	-	17	31	10	42	100
5 Ghatlodiya	i)	Core village	29	71	-	-	-	100
	ii)	Housing Soc.	2	25	50	6	7	100
6 Thaltej	i)	Core village	20	80	-	-	-	100
	ii)	Housing Soc.	-	19	44	10	27	100
7 Chandlodiya	i)	Core village	75	25	-	-	-	100
	ii)	Hoousing Soc.	15	70	15	-	-	100
8 Ranip	i)	Core village	43	32	5	-	-	100
	ii)	Housing Soc.	1	41	56	-	2	100
9 Total Study Area	i)	Core village	30	62	8	-	-	100
	ii)	Housing Soc.	3	23	39	2	33	100

Contd.

APPENDIX - 4 :'G' (CONTD.)

TABLE 32: Criteria for Mate Selection in Children's Marriage - First Preference

Name of Maujas		Variables / Type of residence	Caste	Wealth (a)	Family Background (b)	Education culture (c)	Total (d)
1 Vejalpur	i)	Core village	100	-	-	-	100
	ii)	Housing Soc.	63	10	5	22	100
2 Jodhpur	i)	Core village	100	-	-	-	100
	ii)	Housing Soc.	52	6	16	26	100
3 Vastrapur	i)	Core village	100		-	-	-100
	ii)	Housing Soc.	36	28	-	36	100
4 Memnagar	i)	Core village	100	-	-	-	100
	ii)	Housing Soc.	67	6	-	27	100
5 Ghatlodiya	i)	Core village	100	-	-	-	100
	ii)	Housing Soc.	75	10	7	8	100
6 Thaltej	i)	Core village	100	-	-	-	100
	ii)	Housing Soc.	80	7	-	13	100
7 Chandlodiya	i)	Core village	100	-	-	-	100
	ii)	Housing Soc.	96	2	-	2	100
8 Ranip	i)	Core village	100	-	-	-	100
	ii)	Housing Soc.	100	-	-	-	100
9 Total Study Area	i)	Core village	100	-	-	-	100
	ii)	Housing Soc.	73	6	6	13	100

Contd.

APPENDIX - 4 : 'G' (CONTD.)

Name of Maujas		Type of residence	TABLE 33: Mate Selection Second Preference — Caste (a)	Wealth (b)	Family back-ground (c)	Education & culture (d)	Total 3	TABLE 34: No. of Diseases Children had — More than 4 (a)	Less than (b)	Total
1 Vejalpur	i)	Core village	-	100	-	-	100	16	84	100
	ii)	Housing Soc.	5	65	13	17	100	11	89	100
2 Jodhpur	i)	Core village	-	80	-	20	100	12	88	100
	ii)	Housing Soc.	1	27	10	62	100	4	96	100
3 Vastrapur	i)	Core village	-	94	6	-	100	8	92	100
	ii)	Housing Soc.	30	30	-	40	100	-	100	100
4 Memnagar	i)	Core village	67	22	11	-	100	-	100	100
	ii)	Housing soc.	18	39	22	21	100	8	92	100
5 Ghatlodiya	i)	Core village	-	91	9	-	100	22	78	100
	ii)	Housing Soc.	18	54	-	28	100	15	85	100
6 Thaltej	i)	Core village	39	36	21	4	100	27	73	100
	ii)	Housing Soc.	25	50	-	25	100	29	71	100
7 Chandlodiya	i)	Core village	72	28	-	-	100	11	89	100
	ii)	Housing soc.	91	-	9	-	100	20	80	100
8 Ranip	i)	Core village	-	96	-	4	100	10	90	100
	ii)	Housing Soc.	-	90	1	9	100	29	71	100
9 Total Study Area	i)	Core village	26	63	8	3	100	13	87	100
	ii)	Housing Soc.	8	58	6	28	100	15	85	100

Contd.

APPENDIX - 4 : 'G' (CONTD.)

TABLE 35: Type of Treatment Given to Children | TABLE 36 Vaccine application

Name of Maujas		Type of residence	No doctor (a)	Unani (b)	Homeopathic (c)	Ayurvedic (d)	Allopathic (e)	Total (%)	% of children whom vaccine was applied
1 Vejalpur	i)	Core village	-	53	-	21	26	100	99
	ii)	Housing Soc.	-	-	3	12	85	100	98
2 Jodhpur	i)	Core village	-	26	-	12	62	100	100
	ii)	Housing Soc.	-	5	-	17	78	100	100
3 Vastrapur	i)	Core village	-	38	-	8	54	100	38
	ii)	Housing Soc.	-	-	-	-	100	100	100
4 Memnagar	i)	Core village	-	57	-	-	43	100	50
	ii)	Housing Soc.	8	15	-	8	69	100	85
5 Ghatlodiya	i)	Core village	-	-	33	-	67	100	88
	ii)	Housing Soc.	-	6	2	10	82	100	90
6 Thaltej	i)	Core village	-	64	-	9	27	100	73
	ii)	Housing Soc.	-	28	-	28	44	100	100
7 Chandlodiya	i)	Core village	-	33	-	11	56	100	11
	ii)	Housing Soc.	-	4	-	8	88	100	16
8 Ranip	i)	Core village	5	20	-	14	61	100	71
	ii)	Housing Soc.	-	4	-	11	85	100	79
9 Total Study Area	i)	Core village	1	32	3	11	53	100	71
	ii)	Housing Soc.	1	4	1	12	82	100	91

APPENDIX - 4 : 'H' - Attitude Towards Savings Banks

Name of Maujas		Type of residence	TABLE 37 Variables	TABLE 38 % of people having savings accounts in N. Banks	TABLE 39 % of people having share certificates etc.	TABLE 40 Number of share certificates		% of people using vault
						Less than 3	More than 3	
1 Vejalpur	i)	Core village		68	47	100	-	10
	ii)	Housing Soc.		97	96	77	23	70
2 Jodhpur	i)	Core village		80	85	77	23	13
	ii)	Housing Soc.		100	100	70	30	72
3 i)	i)	Core village		37	5	71	29	-
	ii)	Housing Soc.		100	100	46	54	54
4 Memnagar	i)	Core village		50	17	100	-	8
	ii)	Housing Soc.		83	90	81	19	41
5 Ghatlodiya	i)	Core village		73	67	100	-	12
	ii)	Housing Soc.		96	81	88	12	10
6 Thaltej	i)	Core village		20	17	100	-	-
	ii)	Housing Soc.		96	81	85	15	24
7 Chandlodiya	i)	Core village		28	11	100	-	-
	ii)	Housing Soc.		92	77	95	5	-
8 Ranip	i)	Core village		48	15	100	-	-
	ii)	Housing Soc.		99	88	88	12	3
9 Total Study Area	i)	Core village		47	32	94	6	5
	ii)	Housing Soc.		96	90	81	9	38

APPENDIX - 4 : 'I' - Knowledge of Ahmedabad City & Suburbs

Name of Maujas	Type of residence	Variables	TABLE 41: Idea about City			TABLE 42: Idea about Suburb		
			No idea (a)	Core city (b)	Munici-pallimit (c)	No idea (a)	Outside core limit (b)	Outside Municipal (c)
1 Vejalpur	i)	Core village	43	30	27	26	60	14
	ii)	Housing Soc.	27	34	39	3	73	24
2 Jodhpur	i)	Core village	67	6	27	13	73	14
	ii)	Housing Soc.	20	18	62	2	51	47
3 Vastrapur	i)	Core village	63	-	37	53	47	-
	ii)	Housing Soc.	46	-	54	-	61	39
4 Memnagar	i)	Core village	42	-	58	25	75	-
	ii)	Housing Soc.	39	-	61	-	72	28
5 Ghatlodiya	i)	Core village	50	12	38	26	74	-
	ii)	Housing Soc.	55	24	21	4	58	18
6 Thaltej	i)	Core village	53	-	47	30	67	3
	ii)	Housing Soc.	12	-	88	8	68	24
7 Chandlodiya	i)	Core village	83	11	6	28	72	-
	ii)	Housing Soc.	69	24	7	4	96	-
8 Ranip	i)	Core village	68	16	16	28	72	-
	ii)	Housing Soc.	52	23	25	3	97	11
9 Total Study Area	i)	Core village	58	28	14	30	64	6
	ii)	Housing Soc.	37	44	19	3	75	22

APPENDIX - 4 : 'J' - The Structure of the House

TABLE 43: **Material of the Wall**

Name of Maujas	Type of residence		Variables etc.	Grass, leaves etc. (a)	Stone, mud (b)	Unburnt brick (c)	Burnt brick (d)	Total (%)
1 Vejalpur	i)	Core village		5	6	6	83	100
	ii)	Housing Soc.		-	-	-	100	100
2 Jodhpur	i)	Core village		-	-	13	87	100
	ii)	Housing Soc.		-	-	-	100	100
3 Vastrapur	i)	Core village		-	22	11	67	100
	ii)	Housing Soc.		-	-	-	100	100
4 Memnagar	i)	Core village		-	-	83	17	100
	ii)	Housing Soc.		-	-	-	100	100
5 Ghatlodiya	i)	Core village		6	27	-	67	100
	ii)	Housing Soc.		-	-	-	100	100
6 Thaltej	i)	Core village		-	12	25	63	100
	ii)	Housing Soc.		-	-	-	100	100
7 Chandlodiya	i)	Core village		-	-	17	83	100
	ii)	Housing Soc.		-	-	-	100	100
8 Ranip	i)	Core village		-	12	12	76	100
	ii)	Housing Soc.		-	-	-	100	100
9 Total Study Area	i)	Core village		2	11	16	71	100
	ii)	Housing Soc.		-	-	-	100	100

Contd.

APPENDIX - 4 : 'J' (CONTD.)

TABLE 44: Material of the Roof

Name of Maujas		Type of residence	Variables etc.	Grass, leaves (a)	Corrugated iron, asbestos (b)	Brick lime (c)	Concrete (d)	Total (%)
1 Vejalpur	i)	Core village		23	10	50	17	100
	ii)	Housing Soc.		-	-	-	10	100
2 Jodhpur	i)	Core village		5	15	40	40	100
	ii)	Housing Soc.		-	15	40	40	100
3 Vastrapur	i)	Core village		37	5	32	26	100
	ii)	Housing Soc.		-	-	-	100	100
4 Memnagar	i)	Core village		17	-	58	25	100
	ii)	Housing Soc.		-	-	-	100	100
5 Ghatlodiya	i)	Core village		-	19	75	6	100
	ii)	Housing Soc.		-	-	-	100	100
6 Thaltej	i)	Core village		23	10	40	27	100
	ii)	Housing Soc.		-	-	-	100	100
7 Chandlodiya	i)	Core village		22	6	72	-	100
	ii)	Housing Soc.		-	-	-	100	100
8 Ranip	i)	Core village		4	4	92	-	100
	ii)	Housing Soc.		-	-	-	100	100
9 Total Study Area	i)	Core village		21	7	45	17	100
	ii)	Housing Soc.			-	-	100	100

Contd.

APPENDIX - 4 : 'J' (CONTD.)

TABLE 45: Number of Rooms

Name of Maujas	Type of residence	Variables	One (a)	Two (b)	Three (c)	Four (d)	More than four (e)	Total (%)
1 Vejalpur	i)	Core village	17	47	23	10	3	100
	ii)	Housing Soc.	3	49	42	4	2	100
2 Jodhpur	i)	Core village	20	40	7	20	13	100
	ii)	Housing Soc.	10	20	53	6	11	100
3 Vastrapur	i)	Core village	47	23	10	10	10	100
	ii)	Housing Soc.	-	31	54	15	-	100
4 Memnagar	i)	Core village	17	25	25	17	16	100
	ii)	Housing Soc.	7	24	37	22	10	100
5 Ghatlodiya	i)	Core village	67	27	3	3	-	100
	ii)	Housing Soc.	17	65	15	2	1	100
6 Thaltej	i)	Core village	46	23	23	4	4	100
	ii)	Housing Soc.	12	36	40	4	8	100
7 Chandlodiya	i)	Core village	44	56	-	-	-	100
	ii)	Housing Soc.	13	74	13	-	-	100
8 Ranip	i)	Core village	52	36	12	-	-	100
	ii)	Housing Soc.	22	24	49	5	-	100
9 Total Study Area	i)	Core village	38	36	15	6	5	100
	ii)	Housing Soc.	11	45	35	4	5	100

Contd.

APPENDIX - 4 : J (CONTD.)

TABLE 46: **Built - up Area in SQ.Yard**

Name of Maujas	Type of residence	Less than 60 (a)	60-79 (b)	80-99 (c)	100-149 (d)	More than 149 (e)	Total (%)
1 Vejalpur	i) Core village	77	20	3	-	-	100
	ii) Housing Soc.	15	60	8	17	-	100
2 Jodhpur	i) Core village	66	34	-	-	-	100
	ii) Housing Soc.	20	38	21	10	11	100
3 Vastrapur	i) Core village	79	11	10	-	-	100
	ii) Housing Soc.	31	31	15	23	1	100
4 Memnagar	i) Core village	41	33	9	17	-	100
	ii) Housing Soc.	29	24	27	20	-	100
5 Ghatlodiya	i) Core village	88	8	2	-	2	100
	ii) Housing Soc.	43	33	20	4	-	100
6 Thaltej	i) Core village	57	32	11	-	-	100
	ii) Housing Soc.	48	36	8	8	-	100
7 Chandlodiya	i) Core village	83	17	-	-	-	100
	ii) Housing Soc.	61	37	2	-	-	100
8 Ranip	i) Core village	80	20	-	-	-	100
	ii) Housing Soc.	37	46	14	3	-	100
9 Total Study Area	i) Core village	77	12	4	4	3	100
	ii) Housing Soc.	34	43	11	10	2	100

APPENDIX 4 : 'K' - Facilities Available in Houses

TABLE 47: Households Having Separate Facilities in their Houses

Name of Maujas	Type residence	Vari-ables	Electri-city (a)	Water tap (b)	Bath-room (c)	Kitchen (d)	Toilet (e)
1 Vejalpur	i)	Core village	100	97	97	17	83
	ii)	Housing Soc.	100	100	98	99	98
2 Jodhpur	i)	Core village	100	93	93	93	93
	ii)	Housing Soc.	100	100	100	100	100
3 Vastrapur	i)	Core village	95	53	16	16	11
	ii)	Housing Soc.	100	100	100	100	100
4 Memnagar	i)	Core village	92	75	50	58	33
	ii)	Housing Soc.	100	100	100	100	100
5 Ghatlodiya	i)	Core village	94	81	44	31	63
	ii)	Housing Soc.	100	100	97	99	97
6 Thaltej	i)	Core village	60	47	20	47	3
	ii)	Housing Soc.	100	100	100	100	100
7 Chandlodiya	i)	Core village	94	28	33	22	11
	ii)	Housing Soc.	100	100	94	99	99
8 Ranip	i)	Core village	60	44	40	-	-
	ii)	Housing Soc.	100	100	100	100	97
9 Total Study Area	i)	Core village	84	64	48	32	34
	ii)	Housing Soc.	100	100	98	99	98

APPENDIX - 4 : 'L' - Value of the House

TABLE 48: **Rent in Rupees (Monthly)**

Name of Maujas	Type of residence	Less than 100 (a)	100-199 (b)	200-299 (c)	300-399 (d)	More than 399 (e)	Total
1 Vejalpur	i) Core village	100	-	-	-	-	100
	ii) Housing Soc.	-	6	42	26	100	
2 Jodhpur	i) Core village	100	-	-	-	-	100
	ii) Housing Soc.	-	6	22	44	28	100
3 Vastrapur	i) Core village	25	75	-	-	-	100
	ii) Housing Soc.	-	6	30	34	30	100
4 Memnagar	i) Core village	100	-	-	-	-	100
	ii) Housing Soc.	-	15	19	65	1	100
5 Ghatlodiya	i) Core village	33	67	-	-	-	100
	ii) Housing Soc.	3	43	21	19	14	100
6 Thaltej	i) Core village	67	33	-	-	-	100
	ii) Housing Soc.	-	-	67	33	-	100
7 Ghandlodiya	i) Core village	50	50	-	-	-	100
	ii) Housing Soc.	-	39	33	28	-	100
8 Ranip	i) Core village	71	29	-	-	-	100
	ii) Housing Soc.	13	37	40	7	3	100
9 Total Study Area	i) Core village	60	35	5	-	-	100
	ii) Housing Soc.	4	26	30	23	17	100

Contd.

APPENDIX - 4 : 'L' (CONTD.)

TABLE 49: **Selling Value (in Thousand Rupees)**

Name of Maujas	Type of residence (Variables)		Less than 60 (a)	60-99 (b)	100-149 (c)	150-299 (d)	More than 299 (e)	Total
1 Vejalpur	i)	Core village	74	24	2	-	-	100
	ii)	Housing Soc.	6	73	13	8	-	100
2 Jodhpur	i)	Core village	33	60	-	-	7	100
	ii)	Housing Soc.	3	41	32	9	15	100
3 Vastrapur	i)	Core village	80	20	-	-	-	100
	ii)	Housing Soc.	20	40	20	20	-	100
4 Memnagar	i)	Core village	49	42	9	-	-	100
	ii)	Housing Soc.	7	54	21	18	-	100
5 Ghatlodiya	i)	Core village	77	8	15	-	-	100
	ii)	Housing Soc.	13	71	13	3	-	100
6 Thaltej	i)	Core village	82	15	3	-	-	100
	ii)	Housing Soc.	14	32	32	22	-	100
7 Chandlodiya	i)	Core village	69	31	-	-	-	100
	ii)	Housing Soc.	36	51	13	-	-	100
8 Ranip	i)	Core village	52	48	-	-	-	100
	ii)	Housing Soc.	14	49	34	3	-	100
9 Total Study Area	i)	Core village	51	38	6	4	1	100
	ii)	Housing Soc.	6	50	28	14	2	100

APPENDIX - 4 : 'M' - Internal Density of the Huose

TABLE 50: Persons Per Room

Name of Maujas	Variables: Type of residence	More than 10 (a)	7-10 (b)	4-6 (c)	1-3 (d)	Less than 1 (e)	Total (%)
1 Vejalpur	i) Core village	-	33	67	-	-	100
	ii) Housing Soc.	-	6	60	34	-	100
2 Jodhpur	i) Core village	-	40	33	20	7	100
	ii) Housing Soc.	-	4	30	62	4	100
3 Vastrapur	i) Core village	5	26	58	11	-	100
	ii) Housing Soc.	-	-	31	61	8	100
4 Memnagar	i) Core village	-	17	66	17	-	100
	ii) Housing Soc.	-	7	55	33	3	100
5 Ghatlodiya	i) Core village	6	19	69	6	-	100
	ii) Housing Soc.	-	4	71	21	4	100
6 Thaltej	i) Core village	-	46	47	7	-	100
	ii) Housing Soc.	-	12	40	46	12	100
7 Chandlodiya	i) Core village	17	28	55	-	-	100
	ii) Housing Soc.	1	14	69	14	2	100
8 Ranip	i) Core village	-	56	44	-	-	100
	ii) Housing Soc.	1	14	69	14	2	100
9 Total Study	i) Core village	3	36	54	6	1	100
	ii) Housing Soc.	1	7	55	34	3	100

Contd.

APPENDIX - 4 : 'M' (CONTD.)

TABLE 51: Persons Per SQ. Yard of Built-up Area

Name of Maujas	Type of residence	Variables	More than .99 (a)	0.99-0.50 (b)	0.20 -0.49 (c)	0.10 -0.19 (d)	Less than 0.10 (e)	Total
1 Vejalpur	i)	Core village	-	10	57	33	-	100
	ii)	Housing Soc.	-	-	3	31	66	100
2 Jodhpur	i)	Core village	-	-	40	40	20	100
	ii)	Housing Soc.	-	-	-	9	91	100
3 Vastrapur	i)	Core village	-	16	37	42	5	100
	ii)	Housing Soc.	-	-	-	8	92	100
4 Memnagar	i)	Core village	-	-	33	33	34	100
	ii)	Housing Soc.	-	2	7	10	81	100
5 Ghatlodiya	i)	Core village	-	7	-	62	31	100
	ii)	Housing Soc.	-	1	2	34	63	100
6 Thaltej	i)	Core village	-	13	37	27	23	100
	ii)	Housing Soc.	-	-	8	20	72	100
7 Chandlodiya	ii)	Core village	-	22	66	6	6	100
	ii)	Housing Soc.	-	-	6	36	58	100
8 Ranip	i)	Core village	-	8	60	32	-	100
	ii)	Housing Soc.	-	1	7	26	66	100
9 Total Study Area	i)	Core village	-	10	44	33	13	100
	ii)	Housing Soc.	1	2	3	27	68	100

APPENDIX - 5

Points given to variables for the calculation of 'Modernity Index'

Factor-I : **Socio-economic Status**

A. Information regarding head of the family

1. Place of origin

a) Same mauja	1
b) Core city	3
c) Other parts of Gujarat	5
d) Anywhere in India	7

2. Level of education

a) Illiterate	0
b) Primary	1
c) Secondary	2
d) Higher secondary and vocational	4
e) Graduate	6
f) Masters degree, Law, Engineer, Doctor	9
g) Ph.D and above	13

3. Level of Occupation

a) Cultivators, agricultural labourers, milkman, practising occupation of own caste, housewife	1
b) Mill worker and peon	3
c) Clerk	5
d) Businessman	7
e) Officers, profession and service	10

B. Family in General

1. Type of Family

a) Joint	1
b) Nuclear	5

2. Total family members

a) More than 9	1
b) 7-9	3
c) 4-6	5
d) Less than 4	7

3. Relation with the head of the family

a) Very distant relation (other than mentioned below)	1
b) Cousin	2
c) Unmarried niece and nephew	3
d) Married brother and sister-in-law	4
e) Married great-grandson and grand daughter-in-law	5
f) Unmarried great-grandchildren	6

g) Married grandson and grand daughter-in-law — 7
h) Unmarried brother and sister — 8
i) Unmarried grandchildren — 9
j) Married son and daughter-in-law — 10
k) Widowed mother — 11
l) Unmarried children — 12
m) Wife — 13

4. Percentage of children (below 5 years) to total population
   a) More than 35 — 0
   b) 26-35 — 10
   c) 21-25 — 20
   d) 16-20 — 30
   e) 11-15 — 50
   f) less than 11 — 60

5. Percentage of single population in the age group of 11-20 in relation to total population of that age group
   a) Less than 60 — 0
   b) 60-69 — 10
   c) 70-79 — 20
   d) 80-89 — 30
   e) 90-99 — 40
   f) 100 — 50

6. Level of education of family members
   a) Illiterate — 0
   b) No schooling, primary student — 1
   c) Secondary student, Primary — 2
   d) Secondary, Higher secondary and vocational training student — 3
   e) Higher Secondary, Vocational — 4
   f) College student — 5
   g) Graduate — 6
   h) Engineering, Medical and University Student — 7
   i) Law, Engineering, Doctor and Masters Degree — 9
   j) ph.D and above — 12

7. Level of occupation of family members
   a) Cultivators, agricultural laboureres, milkman, practising the occupation of own caste, housewife, unemployed, retired — 1
   b) Student — 2
   c) Mill worker and peon — 3
   d) Clerk — 5
   e) Businessman — 7
   f) Officer, Profession, Service — 10

C. Female participation

1. Percentage of female student in the age group of 5-20 to total population of that group

a) Less than 60	10
b) 60-69	20
c) 70-79	30
d) 80-89	40
e) 90-99	50
f) 100	60

2. Level of occupation of female (more than 20 years) to total female population

a) Housewife, unemployed maid-servant	1
b) Clerk	3
c) Profession, managers	5

3. Percentage of female (more than 5 years) literacy to total female population

a) Less than 60	10
b) 60-69	20
c) 70-79	30
d) 80-89	50
e) More than 89	60

D. Information regarding Socio-cultural economic background of total family

1. Food Habit

a) Vegetarian	1
b) Egg permitted	3
c) Non-Vegetarian	5

2. Religion

a) Hindu and Jain	1
b) Others	5

3. Caste (concentration of single caste)

a) More than 50%	10
b) 26-50	30
c) Less than 26	50

4. Mother tongue

a) Gujarati	1
b) Language of neighbouring state Marathi, Marwar, Sindhi, etc.)	2
c) Hindi	5
d) Other Indian Languages	10

5. Total monthly income (in Rupees)

a) Less than 500	1
b) 500-999	2
c) 1000-1499	4
d) 1500-1999	6

e) 2000 and above 8

Factor-II: Behaviour-Life-style

A. Possession and use of various modern goods (Secarcity method)

1. Possession
   a) Transistor/Radio 4
   b) Kerosene stove 5
   c) Cycle 25
   d) Wall clock/Table clock 27
   e) Pressure cooker 38
   f) Gas stove 56
   g) Scooter 65
   h) Tape recorder 69
   i) Mixer 73
   j) Dining Table 75
   k) Refrigerator 76
   l) Telephone 91
   m) Car and record player 95
   n) Television 97
   o) Cooking range 98

2. Use
   a) Washing soap 2
   b) Bathing soap 5
   c) Safety rezor 15
   d) Utensil cleaning soap 24
   e) Tooth paste 35
   f) Shampoo 79

B.Reading and recreating habit

1. Newspaper
   a) No 1
   b) Yes 5

2. Language of paper (percentage to total 'yes'
   a) Mother tongue 1
   b) Hindi 3
   c) English 8

3. Magazine
   a) No 1
   b) Yes 5

4. Type of Magazine (% to 'yes')
   a) Religious 1
   b) Film and sports 2
   c) General 3
   d) Children 4
   e) Science & education 5

5. Language of magazine (% to 'yes')

a) Mother tongue	1
b) Any other Indian language	3
c) English	5

6. Library membership

a) No	1
b) Yes	5

7. Members of number of libraries (% to 'yes')

a) Less than three	1
b) More than two	5

8. Type of library (% to 'yes')

a) Circulating	1
b) College/Institutional	3
c) Library having books on various subjects	5

9. Member of the club

a) No	1
b) Yes	5

10 Type of club

a) Local, regional, original pol club	1
b) Any sports club	3
c) High status club	5

C. Behaviour in raising of children 1 & 2 Expected degree for male and female child

a) No expectation, literate, primary	1
b) School leaving	3
c) Graduate, any occupational training	5
d) Post graduate and equivalent degree	7
e) Not fixed (as much as child like to have)	9

3. Criteria for mate selection for children

	Preference Ist	2nd
a) Caste	2	1
b) Wealth	4	2
c) Family background	6	3
d) Education and culture	8	4
		Points

4. Number of diseases of children below five in last four year

a) More than three	1
b) Less than four	5

5. Type of disease

a) Each type	1

6. Treatment for children

a) No doctor 1
b) Unani 2
c) Homeopathic 3
d) Ayurvedic 4
e) Allopathic 5

7. Vaccine applied to children
a) No 1
b) Yes 5

8. Number of vaccine applied
a) One 1
b) Two 2
c) More than two 3

D. Attitude towards banks and savings

1. Saving account in nationalised bank
a) No. 1
b) Yes 5

2. Possession of shares, certificates, etc.
a) No 1
b) Yes 5

3. Number of possession of various types of shares, certificates etc.
a) Less than three 1
b) More than two 5

4. Keeping jewellery in vault
a) No 1
b) Yes 5

E. Knowledge about city and suburbs

1. City
a) No idea 1
b) Core city 3
c) Municipal limit 5

2. Suburbs
a) No idea 1
b) Area outside core city 3
c) Area outside minicipal limit 5

Factor-III: Housing Characteristics

A. Housing Structure
1. Material of wall
a) Grass, leaves 1
b) Stone, mud 2
c) Unburnt brick 3

	d) Burnt brick	4
2.	Meterial of roff	
	a) Grass, leaves	1
	b) Corrugated iron, asbestos	2
	c) Brick and lime, tiles	3
	c) Concrete	4
3.	Number of rooms	
	a) One	1
	b) Two	2
	c) Three	3
	d) Four	5
	e) More than four	8
4.	Built - up area in sq. yard	
	a) Less than 60	1
	b) 60-79	2
	c) 80-99	4
	d) 100-149	7
	e) More than 149	10

B. Facility available in house (scarcity method)

a) Separate electricity	5
b) Separate water tap	9
c) Separate bathroom	12
d) Separate kitchen	15
e) Separate toilet	16

C. Value of the house

1.	Rent in rupees (monthly)	
	a) Less than 100	1
	b) 100-199	2
	c) 200-299	3
	d) 300-399	4
	e) More than 399	7
2.	Selling value (in thousand rupees)	
	a) Less than 60	1
	b) 60-99	2
	c) 100-149	3
	d) 150-299	5
	e) More than 299	8

D. Internal density

1.	Persons per room	
	a) More than 10	1
	b) 7-10	2
	c) 3-6	3
	d) 1-3	5

e) Less than 1	8

2. Persons per square yard built-up area

a) More than .99	1
b) .50 - .99	2
c) .20 - .49	3
d) .10 - .19	5
e) Less than .10	8

Points obtained in Scarcity Method 100 - total percentage of a variable in the total study area.

APPENDIX - 6 : **Calculation of 'Modernity Index'**

Factors / Name of Maujas	Socio-economic Status			Behaviour - Life-style			Housing Structure			Total pattern of suburbs
	Point	Index	Pattern	Point	Index	Pattern	Point	Index	Pattern	
1. Vajalpur	4947	103	M	42769	116	M	8126	104	M	Modern
2. Jodhpur	5466	114	M	52238	141	M	8146	104	M	Modern
3. Vastrapur	4908	103	M	36098	96	T	6975	89	T	Mix
4. Memnagar	4655	97	T	33038	89	T	7913	101	M	Mix
5. Ghatlodiya	5070	106	M	34446	93	T	8053	103	M	Mix
6. Thaltej	4080	85	T	24501	66	T	5915	76	T	Traditional
7. Chandlodiya	4247	89	T	21544	58	T	6994	89	T	Traditional
8. Ranip	4538	95	T	26555	72	T	7496	96	T	Traditional
Total study area	4784	100		37009	100		7825	100		

$$\text{Index} = \frac{\text{Point of a mauja}}{\text{Point of the study area}} \times 100$$

Pattern of each factor : M = Modern (Index more than 100)
T = Traditional (Index less than 100\0

Total Pattern: Modern = Three factors have modern index
Mix = One or two factors have modern index
Traditional = No factor his modern index

Source: Field Survey by the author

APPENDIX - 7

Points given to various amenities (Scarcity method)

1	Balmandir	1
2	Primary School	1
3	Private Dispensary	1
4	Tample	1
5	Post Office	2
6	Bank,	2
7	Secondary School	3
8	Choro	4
9	Hospital	5
10	Maternity Home	5
11	Police station	6
12	Other medical facilities	6
13	Cinema	6
14	Dharamsala	6
15	Church	6
16	Mosque	7
17	Higher secondary school	8
18	Other educational institute	8
19	Higher status club	8
20	Crematorium	8

Methods of giving points = 1 + (T.M. -A.M.)

where, T.M. = Total number of maujas

A.M. = Number of maujas where facilities are available.

APPENDIX - 8

Points for calculation of 'Perception Index' of residents and investigator

A. Perception of the area:

a) Very bad	1
b) Bad	2
c) Fair	3
d) Good	5
e) Very good	8

SELECTED BIBLIOGRAPHY

Abler, R; Adams, J.S; Gould, P.: *Spatial Organisation - The Geographer's View of the World.* Prentics - Hall, Inc. Englewood Cliffs, New Jersey.

Abu-Luchod, J.L.: Testing the Theory of Social Area Analysis: The Ecology of Cairo, Egypt, *American Sociological Review,* 34, 181 - 212, 1969.

Ahmedabad Municipal Corporation : Integrated Urban Development Programme - Ahmedabad 1976-81, published in 1964.

Alam, S.M. & Khan W. : *Metropolitan Hyderabad and its Region: Strategy for Development,* Asia Publishers, Bombay, 1972.

Alonso, W. : "The Theory of the Urban Land Market" Paper and *Proceedings of Regional Science Association,* 6, PP. 149-157,1960.

Alonso, W. : *Location and Landuse: Towards a General Theory of Land Rent,* Cambridge (Mass): Harvard University Press, 1964.

Alonso, W. : *Transportation and Urban Land,* Washington, 1964.

Anderson, T.R: and Bean, L.L. : *The Shevky - Bell Social Area: Confirmation of Results and Interpretation of Social Forces,* 40, 119-124, 1961.

Anderson. T.R: and Colleer, J. : Metropolitan Dominance and the Rural Hinterland, *Rural Sociology,* Vol 21, pp. 152-57, 1956.

Andrew. W.H: and Eshleman, J.R. : The New Community : Characteristics of Migrant and Non-Migrant Residents in the Rural Urban Fringe of a Metropolitan Area in *Ohio, Ohio Agricultural Exp. Research Bulletin.* pp. 9-29, 1961.

Andrews, R.B: : Elements of the Urban Fringe Pattern, *Journal of Land and Public Utility Economics,* 18, 169-183, 1942.

Andrews. W.A. : *A Guide to Urban Studies,* Prentice - Hall of Canada Ltd.

Appleyard, D; Lynch, K; and Myer. J. : *The view from the road.* M.I.T. Press, Cambridge, Mass., 1964.

Bal, S. : The Rural Urban Fringe: A Case Study of Dighori. *The Deccan Geographer,* Secunderabad, Vol. 13. 1975.

Basset. K: and Short, J.R. : *Housing and Residential Structure,* Routledge & Kegan Paul London, 1980.

Beegle, J.A. : Characteristics of Michigan's Fringe Population, *Rural Sociology,* PP. 154-63. 1947. Berry, B.J.L; and Herton, F.E. : Concepts in Social Space, *Geographical perspectives on urban systems.* Englewood Cliffs, N.J., Prentice-Hall, 1970.

Berry, B.J.L; and Horton, F.E. : *Geographic Perspective on Urban Suystem with Integrated system,* Prentice - Hall, Inc. Englewood Cliffs, New Jersey, 1970.

Berry, B.J.L; and Marble, D.F. : *Spatial Analysis: A reader in Statistical Geography,* Prentics - Hall Inc., Englewood Cliffs, New Jersey, 1968.

Berry, B.J.L: and Rees, P.H. : The Factorial Ecology of Calcutta, *American Journal of Sociology,* 74, 445-491, 1969.

Beshers, J.M. : *Urban Social Structure,* Free Press Glencos, 111, 1962.

Bhattacharya, M. : *Administrating the Urban Fringe,* World Press. Calcutta, 1978.

Boat, F.W. : Urban Growth and Land Value Patterns: Government Influences, *Profession Geographer,* Vol.22, 1970, PP. 79-82, 1970.

Bombay Land Revenue Code, 1879.

Bombay Tenancy and Agricultural Lands Act, 1948.

Bora Godwa, P; and Mahadev, C.D. : Changing Nature of Agriculture in the Rural Urban Fringe: A Case study of Bangalore, *Indian Geographical Journal,* Vol. 52, 1977.

Bore Godwa, P; and Mahadev, C.D. : Spatial Pattern of Cooperative Sector Housing Development and Its Role in Urban Fringe Expansion. Transactions of the Institute of Indian Geographers, 1981.

Bose, A. (Ed.) : *Studies in India's Urbanisation 1901-1971,* 1973.

Bourne, L.S. : *Internal Structure of the City-readings on space and environment,* Oxford University Press, London 1971.

Bourne, L.S. : *The Geography of Housing,* Edward Arnold, London, 1981.

Boyl, H. : One Hundred Years of Land Value in Chicago, Chicago: University of Chicago Press, 1953.

Breese, G. : Urban Development problems in India, *A.A.A.G.* Vol. 53, No.3 September 1963,

Breese, G. : "Urbanization in Newly Developing Countries", Bureau of Urban Research, N.J.D.N. Prentice Hall, 1966.

Bricham, E.F. : The Determinants of Residential Land Values, *Land Economics,* Vol.41, PP . 325-34, 1965.

Brookfield, H.C. : On the Environment as perceived, in Board C. et. al (eds), Progress in Geography I, London, Edward Arnold, 1969.

Brush, J.E. : Application of a Model to the Analysis of Population Distribution in Indian Cities, Urban Geography of Developing Countries, R.L. Singh Varanasi (Ed.) *National Geographical Society of India,* Banaras Hindu University, PP . 28-49, 1973.

Brush. J.E. : "Morphology of Indian Cities", India's urban future, Roy Turner (E.D.) Berkeley and Los Angeles: University of California Press, 1962.

Brush, J.E. : The Spatial Structure of Cities in India: Pre Industrial or Post Colonial, *International Geography,* W.P. Adam, F.M.Helleiner (Eds.) Montreal, Canada: 22nd International Geographical Congress, 1972.

Brush. J.E. : "Metropolitan Growth and Spatial Changes in Population Density in India : 1961-71." Moscow 23rd International Geographical Congress, 1976.

Brush. J.E. : "Spatial Patterns of Population in Indian Cities", *Geographical Review,* 58, PP. 362-91.

Burgess, E.W. : The Growth of the City: An Introduction to a Research Project, R.E. Park and E.W. Burgess (Eds) 1925: The city, 47, Chicago.

Carr, S: and Schissler, D. : The City as a trip: Perceptual Selection and Memory in the View from the Road, *Environment and Behaviour,* I 1969.

Carver, H. : *Cities in the Suburb,* Canada University of Toronto Press, 1902.

Chang, Sen-Don : Peking: The Growing Metropolis of Communist China, *Geographical Review,* Vol. 55, No. 3, July 1965.

Cherunilum, F. : Urbanisation in developing countries (A Socio-economic and demographic analysis), Himalaya publishing House, Bombay, 1984.

Chisholmm, M. : Rural Settlement and Landuse, London, Hutchinson University Library, 1970.

Chorley, R.J: and Hagget, P. : Quantitative Techniques in Urban Social Geography, Frontiers In Geographical Tcaching, London, Mathuen and Co. Ltd., 1965.

Clark, Colin : *'Population Growth and Land Use'* Macmillan, London, 1968.

Clawson, M. : "Urban Sprawl and Speculation in Suburban Land", *Land Economics,* Vol. 38, PP. 99-111, 1962.

Cochran, W.G. (Ed.) : *Sampling Techniques,* New York: John Wiley & Sons Ltd., 1968.

Cole, J.P: and King, C.A.M. : *Ouantitative Geography,* London, John Willey and Sons Ltd., 1968.

Conlon, F.F. : *Urbanism and Indian Society: The Aspect of Caste,* A.G. Noble and A.K. Dutta (ED), Tata McGraw Hill Publishing Co. Ltd., New Delhi, 1977.

Cox, K.R. : Conflict, *Power and Politics in the City: A Geographic View,* McGraw-Hill, New York, 1973.

Craik, K.H. : The Environmental Psychology, in Craik, K.H. et. al. *New Directions in Psychology,* Vol. 4, New York: Holt, Rinehart and Winston, 1970.

Davies, W.K.D. : Alternative Factorial Solutions and Urban Social Structure: A Data Analysis Exploration of Calgary in 1971, *Canadian Geographer,* 22, 273 - 297, 1978.

Davis, K. : *The Population of India and Pakistan,* Princeton: N.J. Princeton University Press, 1951.

Desai, Anjana :'Some Aspects of Environmental Perception in the Core City of Ahmedabad', The Research Project, funded by the Indian Council of Social Science Research for a period of 1979-1981.

Desai, Anjana : The Environmental Perception of an Urban Landscape: The Case of Ahmedabad, Ekistics, Vol. 47, No. 283, 1980.

Deshpande, C.D : The Suburbs of Greater Bombay, Occasional Paper No.2, Centre for Urban Research, Osmania University, Hyderabad, 1975.

Dixon, J.W.: and Massey Jr., F.J. : *Introduction to Statistical Analysis,* New York: McGraw Bill Book Company, 1967.

Downs, Roger M.: and Sica, David (eds.) : Image and Environment: Cognitive Mapping and Spatial Behaviour, Chicago: Aldine, 1973.

Duncan, O.D.: and Duncan, B. : Residential Distribution and Occupational Stratification, *American Journal of Sociology* 60, 493-503, 1955.

Duncan O.D., and Roi, A.J. : *Social Characteristics of Urban and Rural Communities,* New York, John Wiley and Sons, 1956.

Fesanmaier, D.R.; Good Child, M.F.; and Morrison, S : "The Spatial Structure of the Rural Urban Fringe - A Multivariate Approach", *The Canadian Geographers* 23 (3), 1979, 255-265.

Fielding, G.J. : *Geography as Social Science,* Harper International Edition, Harper & Fow, Publishers, New York, 1976.

Fine, J.; Gienn, N.D.; and Monte, J.k. : The residential Segregation of Occupational Groups in Central Cities and Suburbs, *Demography,* 8, 91-102, 1971.

Fisher, R.A. : Statistical Method for Research Workers, Edinburgh, Oliver and Boyd, 1958.

Ganguly, M. : "The Nature of the Fringe Areas of Calcutta", *Geographical Review of India,* 24, 59, 1967.

Gibbs, J.P. (Ed.) : *Urban Research Methods,* New York D.Van Nostrand Company, East-West Press, New Delhi, 1966.

Gillion, K. : *Ahmedabad: A Study in Indian Urban History* Berkeley University of California Press, 1968.

Goheen, P.G. : *Victorian Toranto 1850-1900 : Patterns and Process of Growth,* University of Chicago, Geography Research Paper 127, 190.

Goldsmith, H.F.: and Lee, S.Y. : "Socio-economic Status within Older and Larger Metropolitan Area:, *Rural Sociology,* PP 207, 1966.

Golledge, R.G. : "Sydney's Metropolitan Fringe - A Study in Urban Rural Relations", *Australian Geographer,* Vol. 7, PP. 243-55 1960.

Gopi, K.N. : Socio-Economic Structure of a Metropolitan Fringe and its Pattern of Interaction with the Central City: *Urbanization in developing countries,* Alam, S.Manzoor and Pokshishvsky. V.V. (Ed) Osmania University 1976.

Gopi, K.N. : *Process of Urban Fringe Development- A Model,* Concept Publishing Company, New Delhi, 1979.

Goryacheva, A.M. et al. : "Interrelation of Factors of Growth in the System of Indian Cities. *Urbanisation in developing countries.* Osmania University, Hyderabad, 1976.

Gregory, S. : *Statistical Methods and Geographers,* London, Longmans Ltd., 1967.

Hagget, P. : *Geography a Modern Synthesis,* Harper and Row Publisher, New York.

Hagget, P. ; *Locational Analysis in Human Geography,* London: Edward Arnold, 1965.

Hagget, P.: and Chorely, R.J. (ed) : "Models, Paradigms and the New Geography", *Models in Geography,* Methuen & Co.Ltd.London, 1967.

Hammad, R.; and Mocullagh, P. : *Quantitative Techniques in geography : An Introduction,* Clarendon Press, Oxford, 1974.

Hatt, P.K. : The Concept of Natural Area, *American Sociological Review,* II, 423- 27, 1946.

Harvey, D. :'Social Processes, Spatial Form and the Redistribution of Real Income in an Urban System', Chirsholm, M (Ed). *Regional Forecasting,* Butterworth, PP.270-300, 1971.

Herbert, D.T. : *Urban Geography: A Social Perspective,* David and Charles, Newton Abbot, 1972.

Herbert, D.T. : and Williams, W.M. : Some New Techniques for Studying Urban Sub-divisions, Applied Geography 2, *Geographical Polonica,* 3, 93-117, 1964.

Herbert, D.T. and De Silva, S.K. : Social Dimensions of a Non-Western City : A Factorial Ecology of Colombo, *Cambria,* 1, 139-153, 1974.

Hawley, A.H.: and Duncan, D.D. : *Social Area Analysis: A Critical Appraisal Land Economics,* 33, 227-245, 1957.

Hurd, R.M. : *Principles of City Land Values,* New York, 1924.

Jadav, R.S. : and Kulkarni, G.S. : "Landuse Planning in the Rural Urban Fringe of Poona", *National Geographic Journal of India,* 13, 158-167, 1967.

Johnson, J.H. (Ed) : Suburban Growth: Geographical Processes at the Edge of the Western City, John Wiley and Sons, 1974.

Johnson, J.H. : *Urban Geography: An Introduction Analysis,* Pergamon Press, Oxford.

Johnston, R.J. : The Population Characteristics of the Urban Fringe: A Review and Example, *Australian and New Zealand Journal of Sociology,* Vol. 22, PP. 19-33, 1966.

Johnston, R.J. : *Urban Residential Pattern,* Bell, London, 1971.

Jones, E. : *The Social Geography of Belfast,* Oxford University Press, Oxford, 1960.

Jones, P.N. : *Ethnic Areas in British Cities. Social Problems and the*

City: Geographical Perspective. D.T. Herbert and D.M. Smith (Eds), Oxford University Press, Oxford PP. 158-185, 1979.

Kabra, K.N. : *Developing an Urban Fringe,* IPA, New Delhi, 1980.

Karan, P.P. : Perception of Environmental Pollution In Chotanagpur Industrial Area, India, *National Geographer,* Vol. 12, No. 2, 1977.

Karnik, G.G. : Urban Land (Ceiling and Regulation) Act 1976 with rules and guidelines.

Kayastha, S.L.; and Yadav, R.P. : " Aspects of Flood Hazard in the Middle Ganga Plain: A case study of selected settlements of Eastern Uttar Pradesh" Paper presented in the Symposium of the IGU Working Group on Perception of the Environment, Sapporo, Japan, 1980.

Klein, H.J. : The Delimitation of the Town Centre in the Image of its Citizens, E.J. Bril (Ed), *Urban Core and Inner City,* University of Leiden, 1967.

Klove, R.C. : *"The Park Ridge - Berrintion area. A Study of Residential Land Patterns and Problems in Suburban Chicago",* Chicago, University of Chicago Press, 1942.

Knos. D. : *Distribution of Land Values in Topenka,* Lawrence, Kansas, 1962.

Konvalev, S.A. : 'Typology of Suburban Zones', *Voprosy Geography,* No. 87, 1971.

Kumar, P. : *Agricultural Changes in Urban Fringe,* Rajesh Publication, New Delhi, 1980.

Kumar, V.K., and Singh, R. : Perception of Air Pollution: A case study of Kanpur City, *Indian Journal of Air Pollution Control,* Vol.2, No.1, Jan. 1979.

Kurtz, R.A., and Flitcher, J.B. : 'Fringe and Suburbs: A Confusion of Concepts'. *Social Forces,* 38, PP. 32-37, 1958.

Lee, T.R. : Urban Neighbourhood as a Socio-Spatial Scheme. *'Human Relations',* Vol. 21, 1968.

Lewis, G.J. : Leickster - Urban Structure and Regional Relationships. N. Pye (Ed) *Leicester and its regions,* Leichester, 1972.

Lowenthal, D. : 'Research in Environmental Perception and Behaviour' in *'Environment and Behaviour',* 1972.

Lynch, K. : *Image of the City.* Cambridge, Mass: MTT Press, 1960.

Mahadev, P.D., and Bore Gowda, P. : "Impact of Metropolitan Centres and Rural Transformation". A case study of villages around Bangalore. *National Geographer,* Vol. 13, No. 2, 1978.

Mahadev, P.D., and Bore Gowda, P. : "Spatial Strategy for the

Development of Metropolitan Fringe Villages - A case study of Bangalore". Paper submitted to Indo-British Seminar on regional development Policies in developing countries, in Madras, Dec. 8-17, Madras, 1979.

Martin, W.T. : "The Rural Urban Fringe : A study of adjustment of residential location", Eugene, Eugene University of Oregon Press, 1953 (Monographs studies in Sociology No. 1)

Mayer, M.H., et al (ed) : *Readingts in Urban Geography,* Central Book Depot, Kanpur, 1967.

Mayers, R.B., and Beagle, J.A. : "Delimitation and Analysis of Rural Urban Fringe", *Applied Anthropology,* Vol. VI, PP. 14-22, 1947.

Mckain, W.C., and Burnight, R.C. : "The Sociological Significance of Rural-Urban Life - From the Rural Point of View", *Rural Sociology,* 18, PP. 109-16, 1953.

Monkhoure, F.J., and Wilkinson, H.P. : *Maps and Diagrams,* London, Methuen and Co. Ltd. 1967.

Moore, G.E.: 'Residential Mobility: In City'. Commission on College Geography Resource paper No. 13, *Association of American Geographers,* Washington, D.C., 1972.

Mukherjee, D. : "The Concept of Fringe and Its Delimitation the case of Orland, Florida, U.S.A.", *Geographical Review of India,* Vol. 25, PP. 45-55, 1963.

Morgan, B.S. : The Segregation of Socio-Economic Groups in Urban Areas : A comparative analysis, *Urban studies,* 12, 47-60, 1975.

Nag, P. : Perception of Technological Hazards in an Urban Environment : The case of power crisis in Calcutta (India). *National Geographical Journal of India,* Vol.XXVI, Parts 3, 4, 1980 (Science Congress Special Volume).

Nangia, S. : *Delhi Metropolitan Region: A Study in Settlement Geography,* New Delhi: K.B. Publications, 1976.

Newton, P.W., and Johnston, R.J. : Residential Area Characteristics and Residential Area Homogeneity, Environment and Planning, A8, 543-552, 1976.

Noble, A.G., and Dutt, A.K. (Eds.) : *Indian Urbanisation and Planning: Vehicles of Modernization.* Tata McGraw Hill Publishing Co. Ltd., New Delhi, 1976.

Oosthwizen, A.J.G. : "Delimitation of Urban Area", Journal of Geography, Vol. 3, PP. 437-450, 1965.

Phadke, V.S., and Sita, K. : "Delineation of the Fringe Area of an

Indian Metropolis - A case study of Bombay". Presented at the Indo-German Seminar on Urban Fringe at Hyderabad, India, 1982.

Poole, N.A., and Boal, F.W. : *Religious Residential Segregation in Belfast in mid 1969 : A multi-level analysis.* Institute of British Geographers, Special Publications, 51-40, 1973.

Proshsky, H.M. et. al (Ed.) : *Environmental Psychology: Man and His Physical Setting New York:* Holt, Rinchart and Winston Inc.

Pryor, R.J. : "Urban Fringe Residence : Motivation and Stratification in Melbourne". *Australian Geographer,* Vol. ii, PP. 148-56, 1964

Pryor, R.J. : "Defining the Rural Urban Fringe". *Social Force,* 47, 202-25, 1968.

Pryor, R.J. : "The Residential Background of Urban Fringe Dwellers" *Australian Geographical Studies,* Vol. 7, PP. 159-66, 1969.

Queen, S.A., and Carpenter, D.B. : "The Sociological Significance of the Rural Urban Fringe: From the Urban point of view", *Rural Sociology* 18, P. 104, 1953.

Radford, J.P. : Race, Residence and Ideology: Charleston, South California in mid nineteenth century, *Journal of Historical Geography,* 47, 475-88, 1976.

Ramachandran, P. : 'Housing Situation in Greater Bombay' Report No.2, Tata Institue of Social Sciences, P.83, 1972.

Rao, M.S.A : "Economic Changes and Rationality on a Fringe Village", *The Economic Weekly,* Vol. 14, 1962.

Rees, P.H. : Residential Patterns in American Cities: 1960 University of Chicago. Geography Research Paper, 189, 1979.

Robson, T.B. : An Ecological Analysis of the Evolution of Residential Areas in Sunderland, *Urban Studies,* 3, 120-142, 1966.

Robson, T.B. : *Urban Growth: An Approach,* Methuen, London, 1973.

Rodehaver, M.W. : "Fringe Settlement - A Two Direction Movement". *Rural Sociology,* Vol.12, 1947.

Rohrer, W.C., and Hirzal, R. : "A Methodological Note on Demographic Analysis of Rural Urban Fringe", *Rural Sociology,* PP. 91-93, 1957.

Saarinen, Thomas F. : Perception of Environment, Commission on College Geography, Resource Paper No. 5, Washington, DC: Association of American Geographers, 1969.

Saha, Durgadas : "Process of Mental Mapping Among

Cartographers", in S.P. Dasgupta (Ed) *'Essays in Thematic Cartography'* N.A.T.M.O., Monograph 2, 1982.

Schimid, A.A. : "Converting Land from Rural to Urban Uses", Washington Resources for future, 1967.

Schnore, L.F. : The Rural Urban Variable : An Urbanite's Perspective. *Rural Sociology,* Vol. 31, No.2, P. 131-143, June 1966.

Schnore Leo. F. : The Functions of Metropolitan Suburbs. *The American Journal of Sociology,* Vol. 61, No. 2, PP. 433-462, March 1956.

Shaw, M. : The Ecology of Social Change, Wolverhampton 1851-71. *Transaction, Institute of British Geographers,* NS 2, 332-48, 1977.

Shaw, M. : 'Reconciling Social and Physical Space', Wolverhampton 1871. *Transactions, Institute of British Geographers.* NS4, 192-213, 1979.

Short, J.R. : Residential Mobility, *Progress in Human Geography,* Vol.2, 1978.

Singh, L.R. : "Changing Landuse pattern in the Urban Fringe of Allahabad, Umarpur Niwar - A Case Study". *Transactions of the Indian Council of Geographers,* No. 38, PP. 32-45.

Singh, R.L. : "The Rural Urban Fringe Studies". *The Geographers,* Vol. 14, PP. 9-22, 1967.

Sinha, M.M.P. : *"The Impact of Urbanization on Landuse in the Rural Urban Fringe - A case study of Patna",* Concept Publishing Co., New Delhi, 1980.

Smith, T.S. : "The Population of Louvisian: Its Composition and Changes", *Louvisian Bulletin* NO. 293, P. 24, 1937.

Srivastava, B., and Ramachandran, R. : "The Rural-Urban Fringe", *The Indian Geographic Journal,* Vol. 49, PP. 1-9, 1974.

Stewart, C.L. : "Land Values as Affected by Roads and Distances", *Journal of Farm Economics,* Vol. 18, PP. 724-35, 1936.

Suttles, G.D. : *The Social Construction of Communities,* University of Chicago Press, Chicago, 1972.

Swarger, J. : Pittsburgh's Residential Pattern in 1815. Annals *American Association of Geographers,* 68, PP. 265-77, 1978.

Sweetser, F.L. : Factor Structure as Ecological Structure in Helsinki and Boston, Acta, *Sociologica,* 8, 205-225, 1965.

Taeuber, K.E., and Taeuber, A.F. : *Negroes in cities : Residential Segregation and Neighbourhood Change,* Aldine, Chicago, 1965.

Taylor, P.J. : *Quantitative Method in Geography: an Introduction to Spatial Analysis.* Houghton Miffin Company. Boston.

Tembe, S.M. : Raipur : A Study in City and Suburbs, The Deccan Geographer, *Secunderabad,* Vol. 13, 1975.

Timms, D.W.G. : *The Urban Mosaic,* Cambridge University Press, London, 1971.

Timms, D.W.G. : Social Basis to Social Areas, Social Areas in Cities, D.T. Herbert and R.J. Johnston (Eds) Wiley, London, 35-55, 1978.

Tunan, Yi-Fu. : Geography, Phenomenology and the Study of Human Nature, *Canadian Geography,* 25, 1971.

Tunan, Yi-Fu. : Topophilia: A study of environmental perceptions, attitude and value, Englewood Cliffs, New Jersey. Prentice-Hall, 1974.

Tunan, Yi-Fu. : Space and Place: Humanistic Perspective, *in Progress in Geography,* Vol.6, London : Edward Arnold, 1974.

Turner, R. (Ed) : *India's Urban Future,* Berkeley, University of California Press, 1960.

Van Aredol, M.D., Camillesi, S.F., and Schmid, C.F. : The Generality of Urban Social Area Indexes, *American Sociological Review,* 23, 277-284, 1958.

Wibberly, G.P. : *"Agriculture and Urban Growth - A study of the competition for Rural Lands",* London, Michel Joseph, 1969.

Wissinik, G.A. : "American Cities in Perspective : with special reference to the Development of their fringe areas". *Social geografische studies* No. 5 (assen, Netherlands, Royal Gorum), 1962.

Wadva, Kiran : *Urban Fringe Land Markets : An Economic Analysis (a case study of Ahmedabad).* Concept Publishing Company, New Delhi. 1982.

Walker, R.A. : The Transformation of Urban Structre in the Nineteenth Century and the Beginning of Suburbanisation. Kr. T. Cox (Ed), *Urbanisation and Conflict in Market Societies.* Maarofa Press, Chicago, 165-212, 1978.

Walker, T.M. : *"Rural Urban Fringe : A Study of adjustment of residential location".* Oregon: University of Oregon Press, 1963.

Wendt, P.F. : Theory of Urban Land Values, *Land Economics,* Vol. 33, PP. 228-40, 1957.

Wahrwin, G.S. : The Rural Urban Fringe, *Economic Geography,* Vol. 18, PP. 217-28, 1942.

Whitehand, J.W.R. : "Fringe Belt : A Neglected Aspect of Urban Geography". *Institute of British Geographer,* No. 41, P. 223, 1967.

Williams, O.P. : *Metropolitan Political Analysis. A Social Access Approach,* Free Press, New York, 1971.

Woodcock, R.G., and Bailey, M.J. : *Quantitative Geography,* Macdonald and Evace, Plymouth.

Writh, L. : *The Ghetto.* The University of Chicago Press, Chicago, 1928.

Yeats, M. : Some Factors Affecting the Spatial Distribution of Chicago Land Values 1910-60, *Economic Geography* 41, PP. 57-70, 1965.

Zimmer, B.G., and Hawley, A.H. : "Approaches to the Solution of Fringe Problems: Preference of residents in the flint Metropolitan area", *Public Administration Review,* 14, PP. 258-68, 1956.

Zorbaugh, H.W. : *The Gold Coast and the slum,* University of Chicago Press, 1929.

INDEX